FLORA ZAMBESIACA

Flora terrarum Zambesii aquis conjunctarum

VOLUME NINE: PART TWO

FLORA ZAMBESIACA

MOZAMBIQUE

MALAWI, ZAMBIA, ZIMBABWE

BOTSWANA

VOLUME NINE: PART TWO

Edited by
G.V. POPE

on behalf of the Editorial Board:

S.J. OWENS
Royal Botanic Gardens, Kew

I. MOREIRA
*Centro de Botânica, Instituto de Investigação
Científica Tropical, Lisboa*

G.V. POPE
Royal Botanic Gardens, Kew

Published by the Royal Botanic Gardens, Kew,
for the Flora Zambesiaca Managing Committee
1997

Typeset at the Royal Botanic Gardens, Kew, by Christine Beard

Printed in Great Britain by
Whitstable Litho Printers Ltd., Whitstable, Kent.

ISBN 1 900347 23 7

CONTENTS

FAMILIES INCLUDED IN VOLUME IX, PART 2

LIST OF NEW NAMES PUBLISHED IN THIS PART

Acknowledgements

The Flora Zambesiaca Managing Committee thanks M.A. Diniz and E. Martins of
the Centro de Botânica, Lisbon, for their valuable help in reading and
commenting on the text.

Due to circumstances beyond our control we have not been able to include family
number 136 (Polygonaceae) in this part. This will now appear in a later part.

137. PODOSTEMACEAE

By C. Cusset

Submerged freshwater herbs, firmly attached to rocks and stones in swift-flowing water or spray of waterfalls, often resembling mosses, liverworts or algae, habit also adapting to depth of water and receding water levels. Plant base usually thalloid, variable in form and bearing endogenous buds on the margins and surface from which flowering shoots and stems arise. Stems simple or with abbreviated side shoots, leafless to ± densely leafy, and/or with reduced leaflets on flowering branches, or stems suppressed. Leaves linear to filiform or reduced and scale-like; linear leaves floating, entire or dichotomously, pinnately or laciniately divided, exstipulate or sometimes with 2 tooth-like stipules; scale-like leaves when present 3-ranked (tristichous) or 2-ranked (distichous), scattered or absent on the stems, and ± densely imbricate on branches and flowering shoots. Flower hermaphrodite, small, actinomorphic or zygomorphic, without a spathella, or with a persistent membranous spathella (2 spathaceous bracts); spathella at first encloses the flower-bud and at anthesis tears irregularly at the apex allowing the pedicel to elongate bearing the flower erect beyond the spathella; flower-bud subtended by 2–3 free protecting bracts, or flower-bud at first inverted (reflexed) within a membranous spathella (also erect within the spathella outside the Flora Zambesiaca area). Perianth of 3 segments (tepals) connate in their lower part, or perianth reduced to 2 minute free filiform structures, lateral to, and shorter than, the androecium, or perianth absent. Stamens hypogynous, 1 or 2 with filaments connate for at least one-third of their length; anthers 2-locular, dehiscing introrsely by a longitudinal slit. Ovary sessile or on a gynophore, globose to ellipsoid, 1–2 or 3-locular with locules of equal or unequal size; locules ellipsoid to fusiform or subglobose, 2-lobed or not, placentation central or axile, bearing numerous anatropous ovules; styles 2 or 3, sessile or subsessile, usually free, variable in shape. Capsule brown, spherical to ellipsoid or fusiform, smooth or adorned with ± wide longitudinal ribs, dehiscing in dry air into 2–3 equal and sometimes caducous valves, or 2 unequal valves of which only the smaller is caducous. Seeds reddish-brown to blackish, minute, slightly flattened-ellipsoid to ovoid, exalbuminous; testa reticulate; embryo straight.

A family of 42(50) genera widely distributed throughout the tropics and subtropics of Africa, Asia, Australia and the Americas, extending slightly into temperate Japan and eastern North America (1 species in rapids of St. Lawrence River, Canada). Most of the species and genera are endemic to small geographic areas, with only one genus, *Tristicha*, world-wide in distribution.

Plants of this family are freshwater herbs, almost always found in swift-flowing, permanent rivers, usually in turbulent waters of cataracts and waterfalls, or in continuous water-spray, often firmly fixed to rocks by a lichen-like thalloid basal part. Submerged and often annual the plants flower as they become exposed to air when water levels recede. Pollination is usually by insects, or by wind. Fruits develop rapidly after pollination. The seed coat is mucilaginous when wet and adheres to the substrate where it germinates.

1. Flower never enclosed within a spathella, but subtended by 2 scarious bracts larger than the scale-like leaves; tepals 3, scarious, connate below; ovary 3-locular ········· 4. **Tristicha**
– Flower enclosed and reflexed within a spathella before anthesis, borne erect on a long-exserted pedicel from the ruptured spathella after anthesis; tepals 2 minute free filiform structures, lateral to, and shorter than, the androecium; ovary 2-locular ··········· 2
2. Capsule smooth, dehiscing into 2 equal caducous valves; leaves and leaf segments usually 1–2 mm wide, linear ····································· 2. **Leiothylax**
– Capsule longitudinally ribbed, dehiscing into 2 persistent or caducous valves; leaves and leaf segments filiform; scale-like leaves absent or present ····················· 3
3. Capsule globose to subglobose, the smaller valve caducous; scale-like leaves absent ·····
··· 3. **Sphaerothylax**
– Capsule ovoid-ellipsoidal, one or both valves caducous; scale-like leaves dentate, numerous and overlapping on the branches bearing the flowers and ± scattered on the stems ·····
··· 1. **Ledermanniella**

1

1. LEDERMANNIELLA Engl.

Ledermanniella Engl., Bot. Jahrb. Syst. **43**: 378 (1909). —C. Cusset in Bull. Mus. Natl. Hist. Nat., B, Adansonia **5**, 4: 361–390 (1983); **6**, 3: 249–278 (1984). —C.D.K. Cook, Aquatic Plant Book: 183 (1990).

Inversodicraeia Engl. ex R.E. Fr., Wiss. Ergebn. Schwed. Rhod.-Kongo-Exped. **1**, 1: 56 (1914), as "*Inversodicraea*".

Monandriella Engl., Bot. Jahrb. Syst. **60**: 457 (1926).

Submerged freshwater herbs; basal part thalloid, foliaceous, ± deeply divided or ribbon-like, bearing endogenous buds on the surface or margins from which shoots arise; shoots with well developed, usually branched stems. Leaves filamentous or reduced and scale-like; filamentous leaves usually divided, densely arranged or scattered; scale-like leaves variable, dentate or lobed or entire, borne throughout the length of the stems and branches or only on the flowering branches. Flowers solitary and apical on branches, also in the angles of the uppermost branches, each flower at first inverted (reflexed) within the unruptured spathella (spathaceous bracts); spathella dehiscing irregularly at the apex, or laterally. Pedicel long-exserted from the spathella after anthesis, bearing the flower erect. Tepals 2, linear or filiform, one on each side of the andropodium. Stamens 1 or 2 (rarely 3), with filaments joined in the lower part for at least one-third of their length, usually longer than the ovary; pollen in monads or diads. Ovary sessile or carried on a ± well developed gynophore, ovoid to ellipsoid, unilocular with central placentation; styles 2, sessile, usually free, of varying length and shape. Capsule ovoid to ellipsoid, usually 8-ribbed, splitting into 2 equal carpels, the suture occurring along 2 of the ribs (except in *L. warmingiana* where ribs are absent) so that each valve appears to be 5-ribbed after dehiscence; the ribs sometimes produced onto the gynophore and then preventing the valves from falling. Seeds small, reddish-brown to blackish, ellipsoid, ± dorsiventrally flattened, with reticulate testa.

An African genus, comprising 44 species. A slightly enlarged generic description, not limited to the species included in the Flora Zambesiaca, has been given.

Plants slender; scale-like leaves laxly scattered on stem and branches, only becoming closely arranged immediately below the flower ·1. *torrei*
Plants more robust; scale-like leaves more numerous on stem and branches, and densely imbricate on the flowering branches ·2. *tenax*

1. **Ledermanniella torrei** C. Cusset in Bull. Mus. Natl. Hist. Nat., B, Adansonia **5**, 4: 386, t. 10 (1983). TAB. **1**, figs. A1–A3. Type: Mozambique, Gurué, Serra Gurué summit, near Namuli Peak, fl. 9.iv.1943, *Torre* 5155 (LISC, holotype).

Thalloid part unknown. Stems 4–10 cm long, branching many times, beset with scattered scale-like leaves which only become more closely arranged immediately below the flowers. Scale-like leaves 2–3 × 0.2–0.3 mm, usually obovate to oblong, the lowermost entire, becoming 3-dentate or sometimes deeply 4–5-lobed; lobes 0.2–1 × 0.1 mm, narrowly triangular, equal or the middle one much longer. Leaves 2–4 cm long, dichotomously or trichotomously divided into filiform segments. Flowers solitary and apical on inflorescence branches, also in the angles of uppermost branches; spathellas 2.5–4 mm long, obovate, tearing irregularly at the tip at anthesis; flowers at first inverted (reflexed) within the spathella, later borne erect on the pedicel outside the spathella. Pedicel elongating to 10–12 mm after anthesis. Tepals 2, filiform, 1 mm long. Stamens 2, filaments joined for almost their entire length; anthers 1 × 0.2–0.3 mm; pollen in monads. Ovary sessile, 1.8 × 1 mm, ellipsoid; styles filiform 0.7–0.8 mm long. Capsule ellipsoid, 8-ribbed; valves not caducous. Seeds 0.5 × 0.3–0.35 mm.

Botswana. N: junction of Linyanti and Zambezi Rivers, fl. vii.1930, *E.M. Young* s.n. (BM).
Mozambique. Z: Gurué, Serra Gurué summit, near Namuli Peak, fr. 20.ix.1944, *Mendonça* 2166 (LISC).
Species known only from the Flora Zambesiaca area.

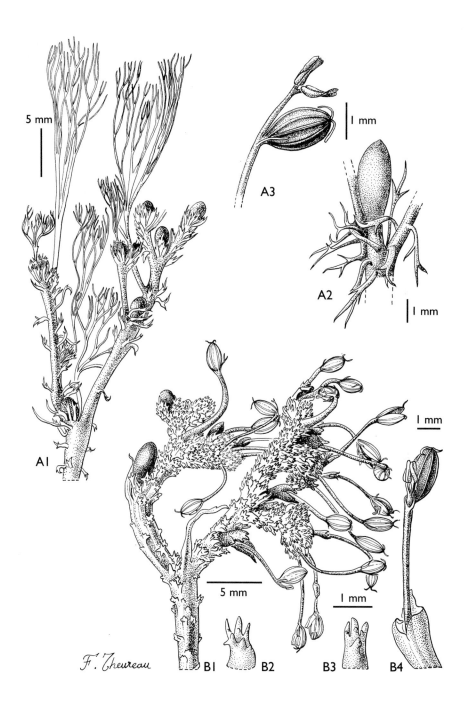

Tab. 1. A. —LEDERMANNIELLA TORREI. A1, branched shoots; A2, flower bud on branched shoot; A3, flower, fully open. B. —LEDERMANNIELLA TENAX. B1, branched shoots; B2,B3, scale-like leaves; B4, flower, fully open. Drawn by F. Theureau.

2. **Ledermanniella tenax** (C.H. Wright) C. Cusset in Adansonia, sér. 2, **14** (2): 275 (1974); in Bull. Mus. Natl. Hist. Nat., B, Adansonia **5**, 4: 384 (1983). TAB. **1**, fig. B1–B4. Type: Zambia, Victoria Falls, Livingstone Island, fl. & fr. 17.ix.1906, *Kolbe* 3149 (K, holotype; BM; BOL).

Dicraeia tenax C.H. Wright in F.T.A. **6**, 1: 125 (1909), as "*Dicraea*".

Inversodicraeia tenax (C.H. Wright) Engl. ex R.E. Fr., Wiss. Ergebn. Schwed. Rhod.-Kongo-Exped. **1**, 1: 56, t. 11, fig. 15–21 (1914). —Engler, Pflanzenw. Afrikas (Veg. Erde 9) **3**, 1: 274 (1915); Bot. Jahrb. Syst. **60**: 462 (1926), as "*Inversodicraea*".

A much-branched submerged perennial herb. Thalloid part ribbon-like, c. 1 mm wide. Stems tough, somewhat fleshy, up to 20 cm long, branched or simple, with scale-like leaves usually scattered on the stem and submerged branches but densely imbricate on the inflorescence branches. Scale-like leaves 1 × 0.5 mm, tridentate with triangular teeth 0.2–0.3 mm long, these leaves becoming longer immediately below the spathellas, increasing to 1.5 × 0.3–0.4 mm with teeth up to 0.7 mm long. Leaves exstipulate, 3–5 cm long, dichotomously or trichotomously divided into filiform segments. Bracts and bracteoles 1.5–2 cm long, dichotomously branched, segments filiform. Flowers 1 or 2 in the angles of the branches and 1–several at the ends of the usually very short uppermost branches; spathellas ellipsoid to obovoid, tearing irregularly at the tip at anthesis, the flowers at first inverted (reflexed) within the spathella, later borne erect on the pedicel outside the ruptured spathella. Pedicels elongating to 6–7 mm after anthesis. Tepals 2, filiform, 0.5 mm long. Stamens 2, filaments c. 1.5 mm long, joined for more than half their length. Anthers 1.5 × 0.3–0.4 mm; pollen in monads. Ovary 2 × 1.2 mm, ellipsoid; gynophore 0.5 mm long; styles filiform, 1–1.3 mm long. Capsule 8-ribbed; valves not caducous. Seeds 0.4 × 0.16 mm, ellipsoid; testa reticulate.

Caprivi Strip. Mambova Rapids, Zambezi R., fl. 21.viii.1958, *West* 3702A (K; SRGH). **Botswana**. N: Chobe R., 2 miles north of Kazane (Kasane), fl. & fr. 13.vii.1937, *Pole Evans* 4214 (K; SRGH). **Zambia**. B: Sesheke, Zambezi R., fr. 17.iv.1969, *Mutimushi* 3741 (K). N: river below the Boma at Mansa (Fort Rosebery), fl. & fr. 6.x.1947, *Brenan* 8026 (K). W: Kabompo R., Solwezi–Mwinilunga road, fr. 23.x.1969, *Drummond & Williamson* 9262 (P; SRGH); S: Zambezi R., 8 miles west of Kazungula (Kasangula), fl. & fr. 25.viii.1947, *Greenway & Brenan* 7983 (COI; K; PRE; SRGH); Victoria Falls, fr. 31.vii.1931, *Fries* 176 (BR; C; K; UPS; Z). **Zimbabwe**. W: Hwange Distr., Victoria Falls, fl. 15.vi.1979, *Mshasha* 205 (K); fl. 21.xi.1949, *Wild* 3129 (K; SRGH). E: Nyanga Distr., Pungwe R., Honde Valley, 11.vii.1965, *Loveridge* 1415 (SRGH).

Also in Angola, Namibia and Tanzania. Locally common, forming colonies densely covering rocks in swiftly flowing water at or just below water level, flowering above water level.

2. LEIOTHYLAX Warm.

Leiothylax Warm. in Kongel. Danske Vidensk. Selsk. Skrift., Naturvidensk. Math. Afd., ser. 6, **9** (2): 147 (1899). —Baker & C.H. Wright in F.T.A. **6**, 1: 124 (1909). —C. Cusset in Adansonia, sér. 2, **20**, 2: 199–207 (1980). —C.D.K. Cook, Aquatic Plant Book: 183 (1990).

Dicraea sect. *Leiocarpodicraea* Engl., Bot. Jahrb. Syst. **20**: 134 (1894).

Leiocarpodicraea (Engl.) Engl., Bot. Jahrb. Syst. **38**: 94, 98 (1907).

Submerged freshwater herbs; basal part thalloid, foliaceous or ribbon-like. Stems long, simple or divaricately branched. Leaves dichotomously branched, segments linear. Spathellas (spathaceous bracts) in clusters along the stems, ovoid to subspherical, rupturing irregularly at the apex at anthesis; flower inverted (reflexed) within the spathella before anthesis. Pedicel long-exserted after anthesis, bearing the flower erect. Tepals 2, minute, one on each side of the andropodium. Stamens 2(3); filaments joined for at least half their length; anthers 2-locular, introrse; pollen in monads. Ovary spherical, 1-locular with central placentation, borne on a well developed gynophore and surmounted by 2 free styles; placentas globular, bearing numerous anatropous ovules. Capsule smooth, globose, dehiscing into 2 equal caducous valves. Seeds numerous, small, red-brown, testa reticulate.

An African genus comprising 3 species of which 2 occur in Cameroon, Zaire and Angola and one in the Flora Zambesiaca area.

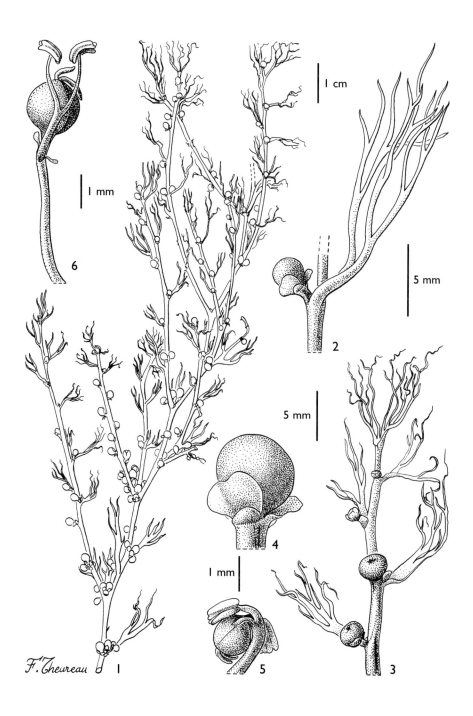

Tab. 2. LEIOTHYLAX DRUMMONDII. 1, habit; 2, flower bud on branched shoot; 3, detail of a branched shoot; 4, flower bud showing spathella; 5, flower with spathella removed; 6, flower, fully open, showing elongated pedicel, tepals, ovary and 2 stamens. Drawn by F. Theureau.

Leiothylax drummondii C. Cusset in Adansonia, sér. 2, **20**, 2: 204, pl. 1 (1980). TAB. **2**. Type: Zambia, Mubalashi R., Kapiri Mposhi–Mkushi road, fl. 8.vi.1963, *Drummond 8271* (P, holotype; SRGH).

Thalloid part foliaceous. Stems ± densely caespitose, up to 50 cm long, branched. Leaves exstipulate, 2–6.5 cm long, dividing dichotomously 2–3 times; leaf segments up to c. 2 mm wide, linear. Spathellas in clusters along the stems, 1–2.5 mm long, subspherical, sessile; flower reflexed within the spathella before anthesis. Pedicel elongating to c. 10 mm after anthesis, bearing the flower erect. Tepals 2, minute, one on each side of the andropodium, 0.3–0.4 mm long. Stamens 2, longer than the ovary; filaments joined for over half their length. Ovary spherical, c. 1.2 mm in diameter, borne on a gynophore 1–1.5 mm long. Styles 2, linear, 0.7–0.8 mm long. Capsule spherical, smooth, dehiscing by 2 equal caducous valves.

Zambia. C: Mkushi Distr., Lunsemfwa R., fast flowing water of hydro canal lined with granite for Kabwe (Broken Hill) Mine, fl. 11.vi.1960, *Mortimer* s.n. (K; SRGH); Lunsemfwa Hydroelectric Plant, canals, fl. 9.vii.1964, *van Rensburg* 2942 (K; SRGH).
So far recorded only from the Lunsemfwa/Mubulashi River system in central Zambia.
H.J. van Rensburg notes that this species was growing submerged in swiftly flowing water along the bottom of the canals of the Lunsemfwa Hydroelectric Plant, where it constituted a problem, being difficult to eradicate.

3. SPHAEROTHYLAX Bisch. ex Krauss

Sphaerothylax Bisch. ex Krauss in Flora **27**: 426 (1844). —C.D.K. Cook, Aquatic Plant Book: 188 (1990).
Anastrophea Wedd. in De Candolle, Prodr. **17**: 78 (1873).

Submerged freshwater herbs; basal thalloid part firmly appressed and attached to substrate, foliaceous and ± deeply lobed or ribbon-like and branched. Stems absent or very short and hardly emerging from the thallus, or elongate and branched. Leaves floating and repeatedly forked into ± filamentous segments. Spathellas (spathaceous bracts with enclosed flower) solitary or aggregated, ± sessile on the thalloid part, and in clusters in the axils of leaves on the elongated stems, subtended by 2 scale-like bracts, spherical to obovoid, dehiscing irregularly at the apex, or laterally; the flower inverted (reflexed) within the unruptured spathella before anthesis. Pedicel long-exserted from the spathella after anthesis, bearing the flower erect. Tepals 2, minute, subulate, one on each side of the andropodium. Stamen 1. Pollen 2-cellular. Ovary 1-locular, subspherical, with central placentation; placentas subglobose bearing numerous anatropous ovules; styles 2, subulate. Capsule subspherical, 8-ribbed, dehiscing into 2 slightly unequal valves, the smaller caducous. Seeds black, numerous, flattened ovate.

A genus comprised of 2 species, one occurring in Madagascar, Angola, Tanzania and Kenya, the other distributed from Zambia to South Africa (Transvaal, KwaZulu-Natal and Cape).

Sphaerothylax algiformis Bisch. ex Krauss in Flora **27**: 426 (1844). —Fries, Wiss. Ergebn. Schwed. Rhod.-Kongo-Exped. **1**, 1: 56, t. 11, fig. 1–14 (1914). —Obermeyer in Fl. Southern Africa **13**: 209 (1970). TAB. **3**. Type from South Africa (KwaZulu-Natal).
Sphaerothylax wageri G. Taylor in J. Bot. **76**: 111 (1938). Type from South Africa (Transvaal).

Thalloid part green or reddish, tightly adhering and appressed to the substrate, foliaceous and ± deeply lobed or ribbon-like, dichotomously branched with branches 1–2 mm wide and, depending on the prevailing conditions (water level), divaricating concentrically to form open lace-like patterns c. 3 cm in diameter. Stems red, erect, 5–50 cm long, simple or branched, ± leafless to densely leafy, at intervals producing side branches consisting of leaves and clusters of few to many bracteate flowers, or stems absent or very short and hardly emerging from the thalloid base. Leaves 1–7 cm long, filiform, simple or repeatedly forked. Spathellas in clusters in leaf axils, and singly and apparently sessile on the lobes and branches of the thalloid base, each subtended by 2 bracts, spherical to obovoid; the flower inverted (reflexed) in the unruptured spathella before anthesis. Pedicel long-exserted after anthesis, up to c.

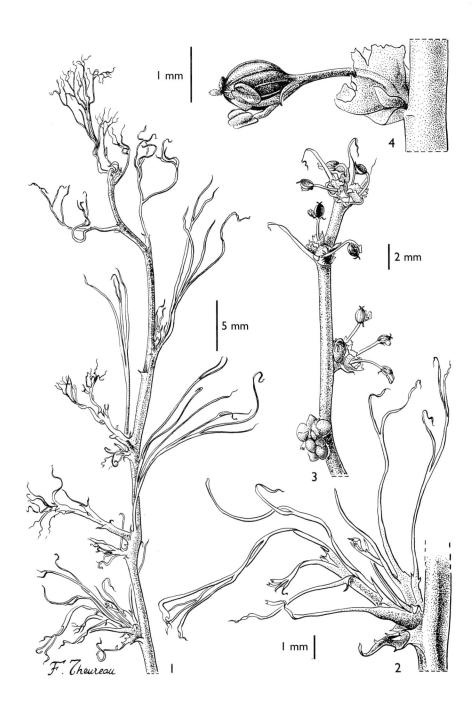

Tab. 3. SPHAEROTHYLAX ALGIFORMIS. 1, upper portion of stem; 2, cluster of leaves before flowering; 3, detail of flowering branch; 4, flower, fully open. Drawn by F. Theureau.

4 mm long, bearing the flower erect from a wide-mouthed funnel-shaped cup. Tepals 2, minute, subulate; stamen 1, filament 1 mm long. Ovary sessile, subglobose, with numerous ovules. Styles small, 0.2 mm long, subulate. Capsule 1.25 × 0.75 mm, subspherical, 8-ribbed, dehiscing into 2 slightly unequal valves, the smaller caducous. Seeds black, numerous, flattened ovate.

Zambia. N: Sansia Falls, Kalambo R., st. 22.i.1961, *E.M. van Zinderen Bakker* 947 (K; PRE). C: Serenje Distr., Kundalila Falls, st. 11.xii.1960, *E.M. van Zinderen Bakker* 944 (K; PRE); north of Serenje, Milenje R., at confluence of Chibamba, st. 22.xii.1963, *Symoens* 10771 (K). S: Zambezi R., 8 miles west of Kasangulu, 25.viii.1947, *Greenway & Brenan* 7982 (K). **Zimbabwe**. C: Harare Distr., Ruwa (Rua) R., st. v.1918, *Eyles* 1321 (BM; SAM; SRGH); fl. vi.1918, *Eyles* 1346 (BM; K; PRE; SAM; SRGH); Ruwa I.C.A., E Harare, Widdecombe Road, fl. 19.v.1963, *Masvanhise* 46 (BM; BR; COI; K; LISC; M; P; PRE; WAG). E: Nyanga Distr., Pungwe R., 16.ix.1964, *Loveridge* 1100 (SRGH); Nyanga, Inyangombe Falls, fl. & fr. 24.i.1960, *Wild* 4884 (BM; K; M; LISC; PRE; SRGH); Nyanga, Gairesi R., fr. 9.ix.1964, *Wild* 6581, 6582 (K; SRGH). **Malawi**. S: Mulanje Distr., Mchese Mt., Fort Lister, fl. 20.vi.1978, *Iwarsson & Ryding* 813 (K); Zomba Distr., Mlunguzi R., b. fl. 4.iv.1969, *Salubeni* 299 (K; LISC; SRGH). **Mozambique**. MS: Gorongosa, fr. 3.vii.1955, *Schelpe* 438 (LISC).

Also in South Africa (Transvaal, KwaZulu-Natal, Cape). Locally common, forming mats on rocks in swiftly flowing water, sometimes in association with other Podostemaceae.

Young stems are usually densely leafy with filiform leaves, the older stems becoming tougher in texture, ± leafless with clusters of flowers usually spaced along their length. When the water level recedes the plants are reduced to the flat thalloid base which often forms ± concentric patterns on the rocks.

4. TRISTICHA Thouars

Tristicha Thouars, Gen. Nov. Madagasc.: 3 (Nov. 1806). —C. & G. Cusset in Bull. Mus. Natl. Hist. Nat., B, Adansonia **10**, 2: 169 (1988). —C.D.K. Cook, Aquatic Plant Book: 188 (1990).
Dufourea Bory ex Willd., Sp. Pl. **5**: 55 (1810).

Submerged, annual or perennial, freshwater herbs, firmly fixed to rock substrate in cataracts and spray at edges of waterfalls. Basal part thalloid, ribbon-like, branched, bearing moss-like shoots which vary greatly from abbreviated flowering shoots to long leafy stems. The abbreviated flowering shoots comprise 2 or more short leafy sterile branches which protect a greatly reduced stem terminated by a solitary flower; the long stems elongating in relation to the depth of submergence, freely branched producing leafy branchlets and short flowering shoots. Leaves small, ± spaced along the stems becoming closely imbricate on the branches and flowering shoots, arranged in 3 ranks (tristichous), sessile, entire or divided. Flowers terminal and/or axillary, each subtended by 2 scarious bracts longer than the leaves; spathella absent. Perianth segments (tepals) 3, joined for part of their length, scarious, usually ovate. Stamen 1. Ovary sessile, 3-locular, topped by 3 free linear styles. Capsule ovoid, ribbed, septicidally 3-valved; seeds small, numerous.

A polymorphic genus of 2 species, one is widespread in tropical and subtropical areas of Central and South America, the West Indies, Africa, Australia, Madagascar, Mauritius and the Mascarene Islands. The other species is native to Australia.

Tristicha trifaria (Bory ex Willd.) Spreng., Syst. Veg., ed. 16, **1**: 22 (1824). —Tulasne in Ann. Sc. Nat., ser. 3, **11**: 111 (1849). —G. Taylor in F.W.T.A., ed. 2, **1**: 124 (1954). —Baker & C.H. Wright in F.T.A. **6**, 1: 121 (1909). —Podlech in Merxmüller, Prodr. Fl. SW. Afrika, fam. 62: 1 (1966). —Obermeyer in Fl. Southern Africa **13**: 206 (1970). TAB. 4. Type from Mauritius.
Dufourea trifaria Bory ex Willd., Sp. Pl. **5**: 55 (1810); in Ges. Naturf. Freunde Berlin Mag. Neuesten Entdeck. Gesamten Naturk. **6**: 63 (1814).

Thalloid part 1–4 mm wide, narrowly ribbon-like, branched, bearing abbreviated flowering shoots and long leafy stems. The abbreviated flowering shoots comprising 2 or more short leafy sterile branches which protect a greatly reduced stem with 2 scarious bracts enclosing an apical flower before anthesis; the long stems up to c. 30 cm long, freely branched, producing leafy branchlets and short flowering shoots.

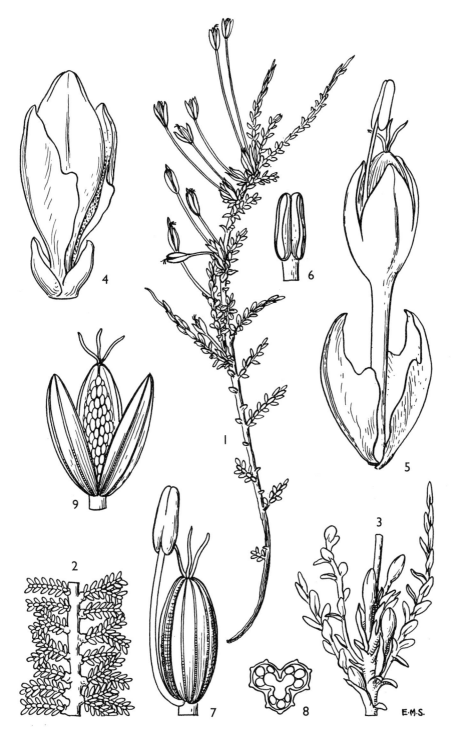

Tab. 4. TRISTICHA TRIFARIA. 1, branched flowering shoot (×3); 2, portion of thallus (×3); 3, portion of flowering shoot (×6); 4, flower bud (×24); 5, mature flower with bracts at base of pedicel (×18); 6, anther (×24); 7, flower with perianth removed (×24); 8, cross section of ovary (diagrammatic); 9, dehisced capsule (×24). Drawn by Margaret Stones. From F.W.T.A.

Leaves ± spaced along the stems becoming closely imbricate on the branches and flowering shoots, arranged in 3 ranks (tristichous), sessile, 0.4–0.6 × 0.5 mm, ovate to ± triangular, obtuse to rounded at the apex, entire or divided, main nerve ± visible. Flowers solitary or in groups, terminal and/or axillary, each subtended by 2 free leafy bracts; bracts longer and thinner in texture than the leaves. Pedicle erect, 2–10 mm long. Tepals 3, scarious, joined for about half their length; free portion 1–2 × 0.5–0.7 mm. Stamen 1 (rarely 2), with a slender filament 2–3 mm long; anther introrse, 0.6–0.7 mm long, oblong. Ovary sessile, 0.7–1 × 0.6 mm, ovoid to ellipsoid, with 3 filiform styles at the apex. Capsule 9-ribbed, dehiscing into 3 equal caducous valves. Seeds orange-brown, small, numerous; testa ornamented by thick closely-spaced anastomosing ribs.

Caprivi Strip. Mambova Rapids, Zambezi R., 21.viii.1958, *West* 3702 (K; SRGH). **Zambia**. N: Kalambo Falls, 22 miles NW of Mbala, fl. & fr. 10.vi.1957, *Seagrief* in *CAH* 2300 (K; SRGH); Sansia Falls, Kalambo R., fr. 22.i.1961, *E.M. van Zinderen Bakker* 948 (K). S: Victoria Falls, Zambezi R., 21.xi.1949, *Wild* 3132 (SRGH). **Zimbabwe**. N: Mazowe Distr., Mazowe (Mazoe) R., st. iv.1907, *Eyles* 544, 545 (BM; SRGH); Makonde Distr., Manyame (Hunyani) Falls, fl. & fr. 13.vii.1921, *Eyles* 3155, 3156 (BM; BOL; SRGH); Murewa Distr., Shavanhohwe (Shawanoe) R., fr. 16.vii.1950, *K. Coates-Palgrave* in *GHS* 28758 (K; SRGH); Mutoko Distr., Mkota Res., Mazowe R., fl. 1.x.1948, *Wild* 2699 (K; SRGH). W: Livingstone Distr., Victoria Falls, fr. 30.viii.1947, *Brenan & Greenway* 7784 (K); Matopos, st. vii.1920, *Eyles* 2544 (BM; PRE; SAM; SRGH); *Eyles* 2545 (BM; K; PRÉ; SAM; SRGH); Shashi R., 8 km above Shashani R. confluence, 7.v.1963, *Drummond* 8118 (P; SRGH). C: Enterprise, Umwindsi R., 19.ix.1947, *Wild* 2002 (SRGH). S: Chiredzi Distr., Chivirira (Chiribira) Falls, Save–Runde confluence, 8.vi.1950, *Wild* 3455 (SRGH); Ndanga Distr., Mutirikwi (Mtilikwe) R., 26.i.1949, *Wild* 2761 (K; SRGH). **Malawi**. C: Naminkokwe (Namakokwi) R., fr. 14.ix.1931, *Burtt Davy* 21693 (K). **Mozambique**. MS: Gorongosa Nat. Park, Nhamusengere Falls, fr. 7.vii.1972, *C.J. Ward* 7780 (K).

Widespread in Africa, extending through West Africa and the Central African Republic to Sudan, Ethiopia, Uganda, Kenya and Tanzania and southwards through Zaire and Angola to Namibia and South Africa (Transvaal, KwaZulu-Natal, and Cape). Also in tropical and subtropical areas of Central and South America, the West Indies, Madagascar, Mauritius and the Mascarene Islands and Australia.

Moss-like and locally abundant forming low reddish-brown mats, attached to rocks at water level (sometimes submerged up to c. 50 cm) in swift-flowing water or in continuous water spray, flowering just above water level.

138. HYDROSTACHYACEAE

By B.L. Stannard

Aquatic herbs attached to rocks in waterfalls and fast-flowing fresh water, usually dioecious, rhizomatous or sometimes stoloniferous. Rhizomes discoid or tuberous; roots numerous, radiating. Leaves submerged or emergent, basal, usually rosulate, often fern-like in appearance with a dorsal (abaxial) and ventral (adaxial) surface, 1–2-pinnate, 2–3-pinnatisect (entire or lobed), dilated and ligulate at the base; petiole and rhachis ± dorso-ventrally flattened and ± densely covered with wart-like or tongue-shaped outgrowths (emergences); pinnae (primary divisions of the leaf) numerous, subopposite, ± cylindrical in outline, bearing simple to variously divided emergences all around the pinna axis, sometimes also with secondary branches (pinnules). Inflorescences emergent, scapiform, densely spicate (resembling fruiting spikes of *Plantago*), produced as water level recedes; peduncle verrucose, rarely smooth or becoming so (on dried specimens at least) as emergences are shed. Flowers unisexual, solitary in the axil of an enclosing bract, sessile; bracts spirally arranged, imbricate, differing in form between the sexes, distal part of bract variously margined or flanged, proximal part or spur [the éperon of Cusset] concave; sepals and petals absent. Male flower reduced to a single subsessile stamen; anther thecae 2, dehiscing longitudinally. Female flower reduced to a 2-carpellate unilocular ovary; ovules numerous, anatropous; styles 2, filiform, exserted from bracts, persistent on the fruit. Fruit a small capsule enclosed by the persistent bract, 2-valved. Seeds small, numerous.

The primary divisions of the leaf, referred to here as pinnae, are called "primary pinnules" in the Flora of Tropical East Africa treatment of the family. The "secondary pinnules" of that Flora treatment are referred to as pinnules in this account.

A monogeneric family found only in Africa and Madagascar. It was previously united with *Podostemaceae* but is now considered to be unrelated.

HYDROSTACHYS Thouars

Hydrostachys Thouars, Gen. Nov. Madagasc.: 2 (1806). —C. Cusset in Adansonia, sér. 2, **13**: 75–119 (1973).

Description as for family.

A genus of 22 species in tropical and southern subtropical Africa, and in Madagascar where a number are endemic. All are found in fast flowing water, usually in cataracts and waterfalls. The species are very variable particularly in leaf form, making them sometimes difficult to distinguish. The changing depth and flow of the water affect plant habit and leaf form. The female bracts appear to be the most constant character.

No material of *H. angustisecta* Engl. has been seen from the Flora Zambesiaca area. The distribution record as given in F.T.E.A. is incorrect, and was probably based on the type locality being given as "Nyassa-See".

Leaves pinnate; pinnae dorsal and ventral surfaces not differentiated, pinna axis beset with filamentous, ± closely spaced, terete, erect-ascending, simple or branched outgrowths (emergences); emergences of rhachis and petiole varying in shape, being simple or branching and linear on the lateral margins, and ovate-lanceolate to triangular on the upper and lower surfaces; apical flange of female bracts faintly 3-lobed · · · · · · 1. *insignis*

Leaves pinnate or bipinnate; pinnae of at least some leaves differentiated into an upper and lower surface, with simple entire ± closely spaced, bract-like emergences on the upper surface, and variously lobed, laciniate or ± filiform-branching emergences on the lower surface, or all pinna emergences simple and bract-like; pinnules (pinna branches) sometimes present on the lower surface; emergences of rhachis and petiole all simple, similar in shape but differing in size; apical flange of female bracts entire · · · 2. *polymorpha*

1. **Hydrostachys insignis** Mildbr. & Reimers in Notizbl. Bot. Gart. Berlin-Dahlem **11**: 662 (1932). —Hess in Ber. Schweiz. Bot. Ges. **63**: 380, t. 9/4 (1953). —Cusset in Adansonia, sér. 2, **13**: 92, t. 2/6 (1973). —Verdcourt in F.T.E.A., Hydrostachyaceae: 2, fig. 1/12–14 (1986). Type from Tanzania.

  Hydrostachys insignis var. *congolana* Hauman in Bull. Soc. Roy. Bot. Belgique **78**: 55 (1946); in F.C.B. **1**: 30, t. 5 (1948). Type from Zaire.

Aquatic caespitose perennial herb from a ± globose woody base, firmly attached to rocks by stiff radiating roots. Leaves fern-like in appearance, few to numerous from rootstock apex, trailing, submerged, up to 55 cm long, narrowly elliptic to lanceolate in outline, pinnate. Petiole 4–20 cm long, covered with ovate, triangular or lanceolate emergences, interspersed with linear or filiform, sometimes branching emergences towards the apex, especially on the lateral margins. Rhachis with numerous ovate to triangular emergences on dorsal and ventral faces, interspersed with lanceolate to linear emergences particularly along lateral margins and towards apex. Pinnae in 55–80 pairs, subopposite to alternate, up to 3.5 cm long, tapering-cylindric to ellipsoid-oblong in outline, bearing branching, filiform emergences all around the pinna axis; filiform emergences 3–5 mm long. Inflorescences scapose, pendulous in bud, stiffly erect at anthesis, few to numerous, spicate. Female inflorescences c. 30–40 cm long; peduncles to 20–30 cm long, covered with emergences; spikes 200–300-flowered; bracts 0.2–0.4 mm long, oblong-elliptic in outline, conchiform with numerous ovate tubercles towards the apex and a flange-like faintly 3-lobed apical appendage. Ovary c. 2 × 1.5 mm, ellipsoid, with a bundle of hairs on each side; styles c. 4 mm long. Male inflorescences up to c. 30 cm long with a dull pale pink axis; peduncles up to c. 20 cm long, densely covered with emergences; bracts 1.5–2.5 × 1.5–2.5 mm, with numerous tubercles towards the apex and a rounded to faintly 3-lobed apical lip, decurrent at the base, pinkish with a dark purple apex. Anther thecae 0.75–1 mm long, elliptic. Fruit not seen.

Malawi. S: Thyolo Distr., Zoa Falls, Ruo R., near Thekerani, c. 450 m, st. 18.iv.1982, *J.D. Chapman* 6135 (K; MAL).

Also in Tanzania, Zaire and Angola. On rocks in cataracts and in rock crevices in waterfalls; 450–1800 m.

The *Chapman* 6135 material agrees well with the type in having at once pinnate leaves with branching filiform emergences on the pinnae, and linear to filiform emergences along the lateral margins of the leaf-rhachis and upper petiole. It does, however, differ from the type in having fewer filiform emergences on the rhachis and pinnae. The significance of this variation cannot be assessed until more material is available.

2. **Hydrostachys polymorpha** Klotzsch ex A. Braun in Peters, Naturw. Reise Mossambique **6**, part 2: 506, tt. 52, 53 (1864). —Weddell in De Candolle, Prodr. **17**: 89 (1873). —Baker & C.H. Wright in F.T.A. **6**, 1: 130 (1909). —Reimers in Notizbl. Bot. Gart. Berlin-Dahlem **11**: 665 (1932). —Hauman in F.C.B. **1**: 29 (1948). —Hess in Ber. Schweiz. Bot. Ges. **63**: 375, t. 11/5 (1953). —Brenan in Mem. New York Bot. Gard. **9**: 58 (1954). —Obermeyer in Fl. Southern Afr. **13**: 213, fig. 32 (1970). —Cusset in Adansonia sér. 2, **13**: 112, t. 2/18 (1973). —Verdcourt in F.T.E.A., Hydrostachyaceae: 3, fig. 1 (1986). TAB. **5**. Type: Mozambique, Moraviland, R. Pômfi (Pomfe), 3 days journey north from Tete, *Peters* s.n. (B, holotype).

Hydrostachys natalensis Wedd. in De Candolle, Prodr. **17**: 88 (1873). —A.W. Hill in F. C. **5**, 1: 484 (1912). Type from South Africa.

Hydrostachys cristulata Wedd. in De Candolle, Prodr. **17**: 89 (1873). Type: Malawi, river flowing into the Shire, north of Chikwawa (Shibisa), *Kirk* s.n. (K, holotype).

Hydrostachys multipinnata Engl., Bot. Jahrb. Syst. **20**: 137 (1894). Type: Malawi, near Blantyre, Shire Highlands, *Last* s.n. (K, holotype).

Aquatic caespitose perennial herb from a ± globose woody base, firmly attached to rocks by stiff radiating roots. Leaves fern-like in appearance, few to numerous from the rootstock apex, trailing, submerged, up to 70 cm long, narrowly elliptic to lanceolate in outline, often 2–3-pinnate, usually with a differentiated upper and lower surface. Petiole and rhachis ± flattened, bearing numerous small ovate to ligulate scale-like emergences; emergences all similar in shape but varying in size and largest on the rhachis, the largest up to 3 × 2 mm; petiole 4–25 cm long. Pinnae in c. 25–70 pairs, alternate to subopposite, 7.5–50 mm long, tapering-cylindric to ellipsoid in outline, those of at least some leaves differentiated into an upper and lower surface, with scale-like emergences all around the pinna axis; emergences simple, entire, ± closely spaced on the upper surface, and variously lobed, laciniate or ± filiform-branched on the lower surface, or all pinna emergences simple and scale-like; pinnules (pinna branches) sometimes present on the lower surface, 4–10 mm long, with linear to elliptic emergences. Inflorescences emergent, few–numerous (up to c. 50), scapose, spicate above, pendulous in bud, stiffly erect at anthesis. Female inflorescences 6–30 cm long; peduncles 3–17 cm long, covered with tiny emergences, (sometimes flaking off on dried specimens to give smooth surface); spikes 150–200-flowered. Bracts of female inflorescence 2–4 × 1.5–2 mm, conchiform, rounded to broadly elliptic in outline, smooth to verruculose, with 1–4 rows of tubercles towards the apex sometimes fused into ridges; the apical lip flange-like, rounded, entire, recurved; the central nerve decurrent below, lateral nerves sometimes prominent. Ovary 1.5–2 × 1–1.5 mm, ellipsoid, glabrous, usually with a bundle of hairs on each side; styles 1–2.0 mm long, glabrous. Male inflorescences 3.5–22 cm long; peduncles 1.5–15 cm long, covered with emergences; bracts 1.5–2 × 1.5–2.0 mm, irregularly shaped, with several rows of tubercles often ± fused into ridges, apical lip entire and rounded or sometimes recurved, the base decurrent. Anther thecae c. 0.5 mm long, elliptic to rounded, joined at base only, divergent. Fruit 2–2.5 × 0.75–1 mm, elliptic, glabrous. Seeds ovoid to oblong.

Zambia. N: Lumangwe Falls, female fl. 16.ix.1958, *Fanshawe* 4848 (K). W: Mwinilunga Distr., Zambezi Rapids, 6 km north of Kalene Hill Mission, male fl. 20.ii.1975, *Williamson & Gassner* 2452 (K; SRGH). E: Luangwa R. Bridge, E side, N of the bridge, st. 5.ix.1947, *Greenway & Brenan* 8048 (K). **Zimbabwe.** E: Mutare Distr., Nyamkwarara (Nyamakwarara) Valley, Stapleford, st. 1.xi.1967, *Mavi* 379 (K; LISC; MO; SRGH). **Malawi.** N: Rumphi Distr., 2 km NE of Rumphi, Chelinda–Rumphi Stream, 1100 m, male fl. 17.viii.1972, *Brummitt & Patel* 12906 (K; SRGH). S: Nswadzi R., east slopes of Thyolo (Cholo) Mt., 840 m, st. 18.ix.1946, *Brass* 17642 (K; SRGH). **Mozambique.** N: Rio Messalo, 23 km along Marrupa–Nungo road, c. 500 m, female fl. 7.viii.1981, *Jansen, de Koning & de Wilde* 107 (K). Z: Quelimane Distr., M'guluni Mission, female fl. Sept., *Faulkner* K66 (K; SRGH). T: Angónia, Rio Livirandzi, 1500 m, st. 3.iii.1980, *Stefanesco & Nyongani* 573 (SRGH). MS: Manica e Sofala at pontoon bridge (19°01'S, 34°10'E), st. 11.xi.1971, *Ward* 7417 (K).

Also in Angola, Zaire, Tanzania, South Africa (KwaZulu-Natal) and Namibia. Locally abundant, sometimes forming large colonies, attached to rocks in swiftly-flowing fresh water, usually in cataracts, rapids and waterfalls, often in mountain streams in forest shade and full sunlight.

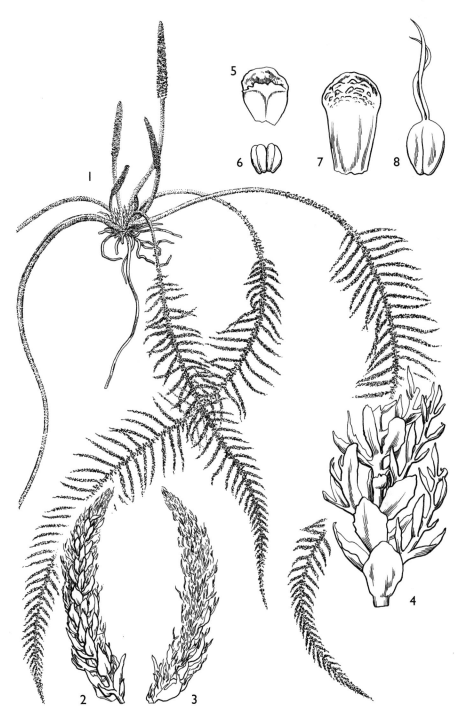

Tab. 5. HYDROSTACHYS POLYMORPHA. 1, habit (×²/₃), from *Baete* in *de Koning* 8277; 2, pinna upper surface showing simple entire emergences (×4); 3, pinna undersurface showing laciniate emergences (×4), 2 & 3 from *Gilliland* 1801; 4, detail of pinna showing pinnules and simple emergences (×6), from *Barbosa* 1825; 5, male bract (×8); 6, anthers (×8), 5 & 6 from *Argosto* 631; 7, female bract (×8); 8, ovary and style (×8), 7 & 8 from *Barbosa* 1825. Drawn by Eleanor Catherine.

Fanshawe 6706 (K; SRGH) from Serenje in Zambia has been placed in *Hydrostachys myriophylla* Hauman by C. Cusset. However, this specimen appears to fall within the range of variation of *Hydrostachys polymorpha* as circumscribed here and for the time being, at least, is not treated as separate.

139. RAFFLESIACEAE

By B.L. Stannard

Parasitic plants on roots and branches of various hosts, thalloid, or vegetatively reduced to an almost mycelial endophytic system with only the flowers emerging from the host tissue, achlorophyllous, monoecious, dioecious, or rarely the flowers of one plant bisexual. Leaves absent (or reduced to scales subtending the flowers). Flowers very small to very large, solitary (rarely in spikes), subtended by bracts, actinomorphic. Perianth ± epigynous, sometimes fleshy; perianth segments (tepals) 3–12, free or basally connate, imbricate or valvate, 1–2-seriate. Stamens 8–many usually fused, in 1–3 series around a fleshy central column; disk sometimes present; anthers sessile, dehiscing by slits or terminal pores. Ovary ± inferior, unilocular; ovules numerous, pendulous from apex of loculus or with parietal placentation. Style 1 or absent. Stigma discoid, capitate or many-lobed. Fruit a berry. Seeds minute, numerous, albuminous.

A tropical and subtropical family of 9 genera and about 50 species in both the Old and New World. The flowers exhibit an extreme range in size, from minute to the largest known flower (up to c. 1 m across), *Rafflesia arnoldii* R. Br. from southeast Asia.

Pilostyles, the most widely distributed genus of the family, is recorded from Iran and Central and South America. Harms (in Pflanzenfam. ed. 2, **16B**: 273 (1935)) included 2 African species in this genus, *Pilostyles aethiopica* Welw. and *P. holtzii* Engl., placing them in section *Berlinianche* on the basis of their having stamens connate into a tube surrounding, and distinct from, the central column (the stamens and stylar tissue are fused into a single column in the rest of *Pilostyles*). I. de Vattimo (in Taxon **4**: 211–212 (1955) subsequently raised section *Berlinianche* to genus level, making the new combinations for the African plants.

BERLINIANCHE (Harms) Vattimo-Gil

Berlinianche (Harms) Vattimo-Gil in Taxon **4**: 212 (1955); in Rodriguesia **26**: 51 (1971). —W. Meijer in Kubitzki et al., Fam. Gen. Vasc. Pl. **2**: 557 (1993).
   Pilostyles sect. *Berlinianche* Harms in Nat. Pflanzenfam. ed. 2: **16B**: 273 (1935).

Small parasites of trees in the family Leguminosae, vegetatively much reduced, only the flowers emerging on the smaller branches of the host. Flowers numerous, involucrate, globose, ovoid or ellipsoid, unisexual by abortion. Perianth segments 4–12, free, similar to the bracts, imbricate, persistent, enclosing an annular disk at the base. Male flowers: stamen filaments connate into a tube about a central column (pistillode); pistillode (and gynandroecial column) expanded apically into a disk (pileus) fringed with papillae; anthers sessile, 1–several-seriate below the pileus rim. Female flowers: ovary inferior surmounted by a fleshy disk; disk flat, concave or convex, sometimes appearing as a simple extension of the style; style short, thick, conical or cylindrical; stigma capitate, the stigmatic surface hemispherical or a subapical annular ring. Berry globose, surrounded by persistent bracts and perianth segments.

A genus of 2 species, confined to tropical Africa.

Berlinianche aethiopica (Welw.) Vattimo-Gil in Taxon **4**: 212 (1955); in Rodriguesia **26**: 51 (1971). TAB. **6**. Type from Angola.
   Pilostyles aethiopica Welw. in Trans. Linn. Soc. London **27**: 67, t. 22 (1869). —Hooker f. in De Candolle, Prodr. **17**: 114 (1873). —Engler, Hochgebirgsfl. Afrika: 201 (1892). —Hiern, Cat. Afr. Pl. Welw. **1**, 4: 908 (1900). —Solms-Laubach in A. Engler, Pflanzenr. [IV, fam. 75] **5**: 15 (1901). —Baker & C.H. Wright in F.T.A. **6**, 1: 131 (1909). —Eyles in Trans. Roy. Soc. South Africa **5**: 345 (1916). —Harms in Engler & Prantl, Pflanzenfam. ed. 2, **16B**: 273 (1935). —Robyns & Boutique in F.C.B. **1**: 391 (1948).

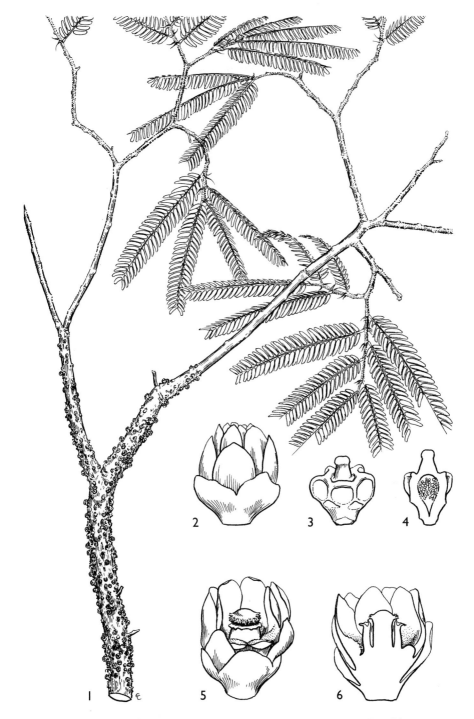

Tab. 6. BERLINIANCHE AETHIOPICA. 1, branchlet of *Brachystegia taxifolius* with emergent flowers of parasite (×²⁄₃); 2, female flower (×8); 3, female flower with perianth segments removed (×8); 4, longitudinal section through female flower (×8), 1–4 from *Sanane* 1221; 5, male flower with 2 perianth segments removed (×8); 6, longitudinal section through male flower (×8), 5 & 6 from *Richards* s.n. Drawn by Eleanor Catherine.

Small internal parasites of trees of *Berlinia, Brachystegia, Isoberlinia* and *Julbernardia* spp., vegetatively much reduced, appearing only as flowers emerging on the smaller branches of the host, pushing through the host's bark. Flowers numerous, usually clustered, sessile, 2–2.5 mm in diameter, hemispherical, bright blood-red turning brown, short-lived, with an unpleasant smell; involucre of imbricate, persistent crustaceous bracts enclosing 1–2 flowers. Involucral bracts 2–2.5 × 2.5–3.0 mm, oblong to rotund, imbricate, persistent. Perianth in 2 whorls of 5–6 segments per whorl; perianth segments purple-red, 1–3.5 × 1.5–2.5 mm, ovate to sub-orbicular or oblong-obovate, obtuse to rounded at the apex, ± fleshy, thickened. Male flowers: staminal tube pink, up to 1 mm tall, separate from and surrounding the pistillode style; anthers white, sessile, 1–3-seriate below the papillate fringe of the umbrella-like pistillode apex; disk white, annular, fleshy, 4–6-angled. Female flowers: ovules numerous, parietal, oblong-ovoid; style to 0.75 mm long, conical; stigmatic surface hemispherical; disk 5–6 angled, slightly concave. Berry globose.

Zambia. N: Mbala Distr., Kawimbe Mission, male fl. 17.vi.1964, *Whellan* 2121 (K; SRGH). C: grounds of Lusaka Hospital, female fl. s.d. *Angus* 3274 (K). S: Mazabuka Distr., Choma to Pemba, mile 26, male fl. 26.iii.1952, *White* 2335 (K). **Zimbabwe.** N: Mutoko Distr., Nyamahere Hill, male fls. 25.ii.1978, *Pope* 1603 (MO; PRE; SRGH). C: Chegutu Distr., Poole Farm, male fls. 6.iii.1962, *H.E. Hornby* in *GHS* 129082 (K; MO; PRE; SRGH). S: Mberengwa, near Mnene Mission, 1931, *Norlindh & Weimarck* s.n. (MO). **Malawi.** C: Lilongwe Distr., Dzalanyama For. Res., male fls. ii.1979, *Alder & Dearden* s.n. (K).

Also in Angola, Zaire and Tanzania. Parasitising species of *Berlinia, Brachystegia, Isoberlinia* and *Julbernardia*; 1000–1676 m.

Herbarium specimens with male flowers appear to be far more common than those with female flowers.

140. HYDNORACEAE

By Lytton J. Musselman

Subterranean parasitic herbs, on roots of shrubs and trees, with a rhizome-like plant-body attached by unbranched haustoria to the host roots, without chlorophyll and leaves. Flowers hermaphrodite, produced endogenously from the "rhizome", large, 3–4(5)-merous; perianth tubular, valvately lobed at the apex. Androecium inserted on the hypanthium, a complex structure composed of numerous fused anthers which form a large undulating ring (or anthers connate and forming a dome in *Prosopanche*), with numerous elongate bisporangiate pollen sacs. Ovary inferior, 3–4-carpellate, unilocular, with numerous infolded, pendant accrescent placentas; ovules numerous. Stigma 4-lobed, sessile and cushion-like. Fruit a subterranean fleshy berry with a woody pericarp; seeds numerous, minute.

A small family comprising 2 genera, *Prosopanche* occurring in Central and South America and *Hydnora* in Africa and Madagascar.

HYDNORA Thunb.

Hydnora Thunb. in Kongl. Vetensk. Acad. Handl. **36**: 69 (1775).
Aphyteia L., Dissert. Planta Aphyteia [respondent E. Acharius]: 7 (1776); Amoen. Acad. **viii**: 310, t. 7 (1785).

Subterranean root parasites. Plant-body a large, verrucose, firmly-fleshy "rhizome" with widely spreading branches. Leaves absent. Flowers arising directly from the plant body and remaining partially below ground level, hermaphrodite, actinomorphic, solitary, 3–4(5)-merous. Perianth fleshy-coriaceous, valvately 3–4(5)-lobed; the lobes equalling or exceeding the perianth tube in length. Androecium inserted on the hypanthium, comprising a complex structure of 3–4(5) fused anthers in the form of a large undulating ring with very numerous horizontally elongate pollen sacs; undulations opposite the perianth lobes. Ovary inferior, unilocular, with numerous infolded, pendant accrescent placentas; ovules numerous.

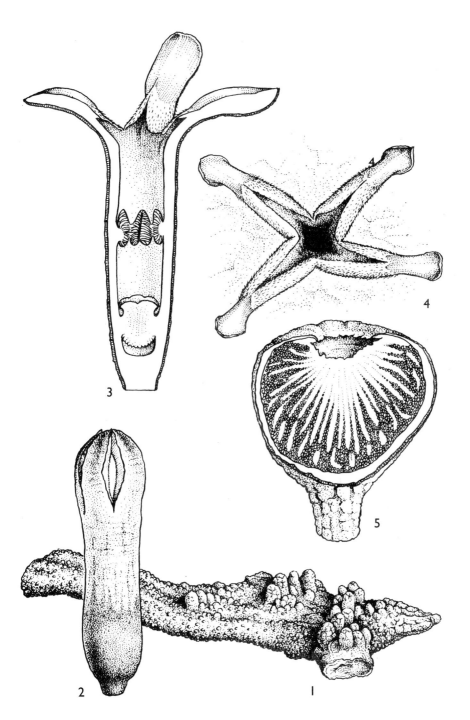

Tab. 7. HYDNORA ABYSSINICA. 1, portion of branching underground plant body ($\times\frac{1}{2}$); 2, flower with perianth lobes beginning to separate ($\times\frac{1}{3}$); 3, longitudinal section through flower showing undulating ring of the androecium and the unilocular inferior ovary ($\times\frac{1}{2}$); 4, perianth lobes spreading above ground level ($\times\frac{1}{2}$); 5, longitudinal section through the subterranean fruit ($\times\frac{1}{3}$). Drawn by Janet Dyer. Reproduced with permission of Notes RBG, Edinburgh.

Stigma sessile, cushion-shaped, 3–5-lobed. Fruit a subterranean fleshy berry with a woody pericarp; seeds numerous in a flesh-coloured pulp, minute.

A genus of c. 12 species. in tropical and South Africa and Madagascar.

Hydnora abyssinica A. Braun ex Schweinf., Beitr. Fl. Aethiop.: 217 (1867). —Engler, Pflanzenw. Ost-Afrikas **C**: 169 (1895); Bot. Jahrb. Syst. **28**: 386 (1900). —Baker & C.H. Wright in F.T.A. **6**, 1: 133 (1909). TAB. **7**. Type from Ethiopia.
Hydnora johannis Becc. in Nuovo Giorn. Bot. Ital. **3**: 5 (1871). —Solms-Laubach in A. Engler, Pflanzenr. [IV, fam. 76] **5**: 7 (1901). —Harms in Engler & Prantl, Pflanzenfam. ed. 2, **16B**: 291 (1935). —Lebrun in F.C.B. **1**: 393 (1948). —Malaisse in Bull. Jard. Bot. Belg. **52**: 115 (1982). —Musselman in Notes Roy. Bot. Gard. Edinburgh **42**: 23, fig. on page 24 (1984); in Dinteria, No. 19: 77 (1987). Type from Ethiopia.
Hydnora bogosensis Becc. in Nuovo Giorn. Bot. Ital. **3**: 6 (1871). Type from Ethiopia.
Hydnora hanningtonii Rendle in J. Bot. **34**: 55 (1896). Type from Kenya.
Hydnora abyssinica var. *quinquefida* Engl., Bot. Jahrb. Syst. **28**: 386 (1900). Type from Tanzania.
Hydnora solmsiana Dinter in Deutsch-Sudw.-Afr.: 57 (1909). —Schreiber in Merxmüller, Prodr. Fl. SW. Afrika, fam. 41: 1 (1968). Type from Namibia.
Hydnora gigantea Chiov., Result. Sc. Miss. Stephan-Paoli Somal. Ital. **1**: 156 (1916). Type from Somalia.
Hydnora africana sensu Eyles in Trans. Roy. Soc. South Africa **5**: 345 (1916).
Hydnora ruspolii Chiov. in Boll. Soc. Bot. Ital. **1917**: 57 (1917). Type from Ethiopia.
Hydnora michaelis Peter in Feddes Repert. Spec. Nov. Regni Veg., Beih. **40** [Peter, Fl. Deutsch-Ostafrika], part 2: 185 (1932). Type from Ethiopia.
Hydnora cornii Vacc. in Nouvo Giorn. Bot. Ital. n.s., **39**: 304 (1932). Type from Somalia.

A subterranean root parasite. The plant-body rhizome-like, verrucose, up to c. 10 cm wide, ± irregular in shape and somewhat flattened, simple or with widely spreading horizontal branches, firmly-fleshy, brick-red inside, containing mucilage ducts, with sticky exudate when fresh, rich in tannins. Branches ± terete and densely covered with small conical warts. Flowers emerging directly from the "rhizome" and branches, remaining partially below ground, solitary or several remotely scattered. Perianth fleshy-coriaceous, 5–25 cm long, tube 3–4 cm in diameter, brown and roughly corky outside, pinkish inside. Perianth lobes usually 4, appearing above ground, fully patent in wet weather and lying on the soil surface, otherwise connivent or even connate at the tip, 6–8 cm long, lanceolate, thickly fleshy, broadly channelled and glandular on the inside; the lobe tips (cucullus) ± thickened, triangular in cross-section or grooved outside, the adjacent cucullus faces glabrous and cushion-like; the margins below the lobe-tips broad and beset with subulate setae. Anthers continuously joined at their bases in a series of large inverted "V"s forming a wavy ring, the anther structure transversally striate and divided into very numerous horizontal pollen sacs; pollen sticky, adhering to the anthers. Ovary unilocular, with numerous infolded, pendant placentas. Stigma sessile, cushion-shaped, 3–4(5)-lobed; each lobe finely ribbed and covered with a brown liquid in living material. Fruit entirely subterranean, fleshy, 10–15 cm in diameter, globose, many-seeded, often splitting irregularly at maturity. Outer layer of fruit a scaly periderm; inner pericarp mealy, white, very sweet to taste. Seeds brown, 1–1.77 mm, oblong to globose. Seedlings unknown.

Botswana. N: Okavango Swamp, *P.A. Smith* s.d. (ODU). **Zimbabwe**. C: Shurugwi Distr., Umcima Farm on Umtebekwa (Tebakwe) R., 2.i.1965, *J. Guy* in *GHS* 159891 (K; SRGH); Gweru Distr., 21.ii.1967, *Biegel* 1929 (K; SRGH). W: Bulawayo, 15.iv.1986, *Musselman & Obilana* s.n. (ODU; M).
Also in the Sudan, Ethiopia, Somalia, Uganda, Kenya, Tanzania, South Africa (Northern Province) and Swaziland. In *Acacia* woodland and scrub, *Acacia-Commiphora* scrub and grassland with scattered *Acacia* spp., usually on black cotton, clay and sandy alluvial soils, also on rocky slopes. Parasitic on *Acacia* spp. but also reported on *Albizia*, *Delonix* and probably *Commiphora* spp.
Plants usually flower after the onset of the rains. Flowers are extremely fetid, visited by beetles. Fruits mature about five months after the flowers appear, are scaly brown and entirely subterranean. Fruit is sweet, with white flesh and abundant brown seeds.
Fruits are frequently collected as food by man and eaten by animals. Other parts of the plant are eaten by animals. Roots are used as medicine for diarrhoea and to staunch haemorrhaging.
In *H. africana* (from Angola and Namibia and the Cape Province in South Africa) the branches are 5–7-angled with warts only on the angles, and the perianth usually 3-lobed with the lobe margins setose to the apex.

141. ARISTOLOCHIACEAE

By B.L. Stannard

Perennial herbs or shrubs, erect, scrambling or climbing, sometimes lianes, often with tuberous rhizomes. Leaves alternate, petiolate, exstipulate, simple, entire, sometimes lobed. Flowers terminal, axillary or borne on the stem (cauligerous), solitary, fasciculate or in few to many-flowered racemose or cymose inflorescences, hermaphrodite, epigynous, actinomorphic or zygomorphic, with or without bracts. Calyx petaloid, usually enlarged, basally tubular, expanding into a limb above; limb symmetrically 3-lobed, or unilateral and entire or lobed. Petals absent except in *Saruma*, vestigial in *Asarum*. Stamens 6–many in 1–2 or more whorls, free or adnate to the stylar column forming a gynostemium; anthers 2-thecous with 4 pollen sacs, extrorse, dehiscing longitudinally. Ovary inferior, rarely half inferior, 4–6 locular, syncarpous or apocarpous in *Saruma*, placentas parietal or seemingly axile; ovules numerous, anatropous, in 1–2 vertical series in each locule, horizontal or pendulous; styles connate into a column; stigma 3–many-lobed. Fruit a capsule, dehiscing septicidally or irregularly, or more rarely indehiscent. Seeds usually many, variously shaped, raphe sometimes thickened and winged; endosperm abundant, embryo minute.

A family of about 12 genera and 450–500 species, mainly in the tropics and subtropics but some in the warm temperate regions. The genus *Pararistolochia* occurs in the F.T.E.A. area, but apparently not in the Flora Zambesiaca region.

ARISTOLOCHIA L.

Aristolochia L., Sp. Pl.: 960 (1753); Gen. Pl., ed. 5: 410 (1754). —H. Huber in Kubitzki et al., Fam. Gen. Vasc. Pl. **2**: 129 (1993).

Perennial herbs or shrubs, erect, scrambling or climbing, often with a fetid smell. Leaves entire or 3–7-lobed, usually cordate, 3–7-nerved, sometimes with an axillary stipule-like leaf. Flowers usually zygomorphic, solitary, fasciculate or in racemose inflorescences, sometimes very large. Perianth coloured, tubular; the tube constricted briefly above its attachment to the ovary apex, then abruptly inflated into a subglobose or oblong-ellipsoid utricle, cylindrical or funnel-shaped above the utricle and straight or variously bent, expanding into a very variable, sometimes unilateral, entire or 3–6-lobed limb. Stamens usually 6, rarely more; anthers sessile, adnate to the stylar column. Ovary inferior, usually merging imperceptibly into the pedicel, usually 6-locular; style with 3–6 stigmatic lobes. Fruits capsular, 6-valved, dehiscing septicidally. Seeds numerous, usually with a peripheral wing, often ± covered on one side by the thickened raphe.

A genus of c. 120 species, widely distributed in the tropics and subtropics with some occurring in the warm temperate regions. Four species are native in the Flora Zambesiaca area. One species, *A. elegans* Mast., introduced from South America, has now become naturalized in some areas. *A. gigantea* Mart. & Zucc., also from South America, characterized by its huge flowers (limb up to 38 × 28 cm) is cultivated in the Flora Zambesiaca region.
   The perianth acts as a trap to attract pollinating insects.
   Dr. C. Neinhuis (Bonn, Germany), who is revising the African species of the genus, was consulted in the preparation of this treatment. His helpful advice is acknowledged here.

1. Flowers large when fully opened, limb 4.5–38 × 5–28 cm, tube sharply reflexed over the inflated basal part, limb peltate (i.e. developed excentrically all round the throat), often crimson with feathery creamy marbling; plants introduced from S America · · · · · · · · · 2
   – Flowers small, limb less than 4 × 2 cm, tube straight, limb consisting of one narrow elongate unilateral lobe, plain dark purplish to crimson-brown or black (no marbling); plants native · 3
2. Flowers huge, limb 13–38 × 9–28 cm; leaves usually slightly longer than broad, broadly ovate; plants cultivated · *gigantea*
   – Flowers smaller, limb 4.5–11 × 5–9 cm; leaves broader than long, broadly ovate-reniform; plants cultivated and naturalized · 4. *elegans*

3. Leaves narrow, linear-lanceolate to narrowly oblong- or ovate-lanceolate · · · · · · 2. *hockii*
- Leaves broader, triangular to ovate · 4
4. Plant usually glabrous, less often leaves minutely puberulous on lower surface; leaf base cordate with deep sinus; flowers 2–9(15) in well developed elongate racemes · · · 1. *albida*
- Plant pubescent; leaf base rounded to subtruncate; flowers usually solitary, more rarely 2(3) in leaf axils or in very short racemes · 3. *heppii*

1. **Aristolochia albida** Duch. in Ann. Sci. Nat., Bot., Sér. 4, **2**: 75 (1854); in De Candolle, Prodr. **15**, 1: 483 (1864). —Hiern, Cat. Afr. Pl. Welw. 1 part IV: 911 (1900). —Baker & C.H. Wright in F.T.A. **6**, 1: 137 (1909). —Peter in Feddes Repert. Spec. Nov. Regni Veg. Beih. **40** [Peter, Fl. Deutsch-Ostafrika], part 2: 184 (1932). —A. Chevalier, Fl. Afr. Occ. Fr. **1**: 131 (1938); in F.W.T.A., ed. 2, **1**: 81 (1954). —Verdcourt in F.T.E.A., Aristolochiaceae: 8 (1986). TAB. **8**, fig. A. Type from Senegal.
   Aristolochia petersiana Klotzsch in Monatsber. Königl. Preuss. Akad. Wiss. Berlin **1859**: 599 (1859); in Peters, Naturw. Reise Mossambique **6**, part 2: 501 (1864). —Engler, Bot. Jahrb. Syst. **24**, t. 10H (1898). —Baker & C.H. Wright in F.T.A. **6**, 1: 137 (1909). —S. Moore in J. Linn. Soc., Bot. **40**: 183 (1911). —Peter in Feddes Repert. Spec. Nov. Regni Veg. Beih. **40** [Peter, Fl. Deutsch-Ostafrika], part 2: 184 (1932). —Hauman in F.C.B. **1**: 383 (1948). —F. White, F.F.N.R.: 44 (1962). —Verdcourt & Trump, Common Pois. Pl. E. Afr.: 18 (1969). — Biegel & Mavi, Rhod. Bot. Dict. Pl. Names [ed. 2 of H. Wild 1953]: 96 (1972). Type: Mozambique, Tete (Tette), Rios de Sena, *Peters* s.n. (B†, holotype; BR; K).
   Aristolochia densivenia Engl., Bot. Jahrb. Syst. **24**: 489, t. 10A–G (1898). —Baker & C.H. Wright in F.T.A. **6**, 1: 136 (1909). —Peter in Feddes Repert. Spec. Nov. Regni Veg. Beih. **40** [Peter, Fl. Deutsch-Ostafrika], part 2: 184 (1932). —Agnew, Upland Kenya Wild Flowers: 84 (1974). Syntypes from Tanzania.
   Aristolochia kirkii Baker in F.T.A. **6**, 1: 139 (1909). Syntypes: Mozambique, Morrumbala (Moramballa), lower Shire River, *Kirk* s.n. (K); Chiramba (Shiramba), lower Zambezi, *Kirk* s.n. (K).
   Aristolochia bainesii Burtt Davy in Bull. Misc. Inform., Kew **1924**: 231 (1924); in Fl. Pl. Ferns Transvaal: 115 (1926). Type: Zimbabwe, "S. African Goldfields", *Baines* s.n. (K, holotype). [Burtt Davy put this locality in the Transvaal. Verdcourt in F.T.E.A. believes that it means only that Baines was working for the S. African Goldfields Exploration Company. Certainly Baines' itinerary would favour Zimbabwe]

Creeping or climbing (twining) perennial herb with annual stems from a woody rootstock, pale green to glaucous. Stems to 3(11) m long, sometimes very slender, sulcate (in dried state), pubescent to glabrous, sometimes tinged purplish at the nodes. Leaves 2–20 × 1.5–15 cm, ovate to ovate-oblong or triangular, rounded to obtuse or acute and ± mucronulate at the apex, cordate or cordate-auriculate with a deep sinus at the base, or leaves 3-lobed with basal lobes strongly patent; terminal lobe much larger than the lateral lobes, 1.5–7(10) × 0.6–3(4) cm, lanceolate to ± ovate; lateral lobes 0.5–2(3) cm in diameter, rotund; lamina glabrous or somewhat puberulous on both surfaces particularly on the veins and around the margins, with 3–7 principal veins from base, the veins and reticulation prominent on the lower surface; petioles 3–40 mm long, pubescent to glabrous; pseudostipules leaf-like, 5–20 × 2.5–14 mm, rotund to ovate or oblong, rounded to subacute sometimes mucronate at the apex, cordate at the base, glabrous or minutely puberulous mainly on veins and margins. Inflorescences axillary, racemose, 2–30(40) cm long, usually elongate, 2–9(15)-flowered; rhachis pubescent to glabrous; bracts conspicuous, persistent, 0.3–4 × 0.2–3 cm, ovate to rotund, rounded to acute and sometimes mucronate at the apex, rounded to cordate and ± amplexicaul at the base, usually glabrous, less often pubescent to minutely puberulous. Flowers, including pedicels and ovary, 0.6–3(4) cm long, glabrous or pubescent, ovary glabrous. Perianth glabrous: constricted part below the utricle 0.5–5 mm long; inflated part (utricle) 2–8 mm in diameter, oblong-ellipsoid to subglobose; tube 2.5–20 × 0.75–2.5 mm, straight, cylindrical, dilated at apex, greyish to purplish-green; limb 8–40 × 3–20 mm, subrotund, oblong-elliptic, elliptic, obovate or panduriform, hairy towards the base, dark purple to purplish-black or brown, sometimes shading to creamy-white at the base. Gynostemium 2–3.5 mm tall, anthers 6. Fruit a capsule, 2–5 × 1–3 cm, oblong-cylindrical to obpyriform, 6-ribbed, apex rounded, base attenuate into a stipe 1–4 cm long; when ripe the pendulous capsule and stipe split longitudinally into 6 segments which remain joined to the pedicel by the thread-like elements of the stipe, the segments separating widely about the middle while remaining joined at the capsule apex, the whole resembling an inverted "parachute". Seeds 6–15 × 4–10 mm, triangular, flat, surrounded by a

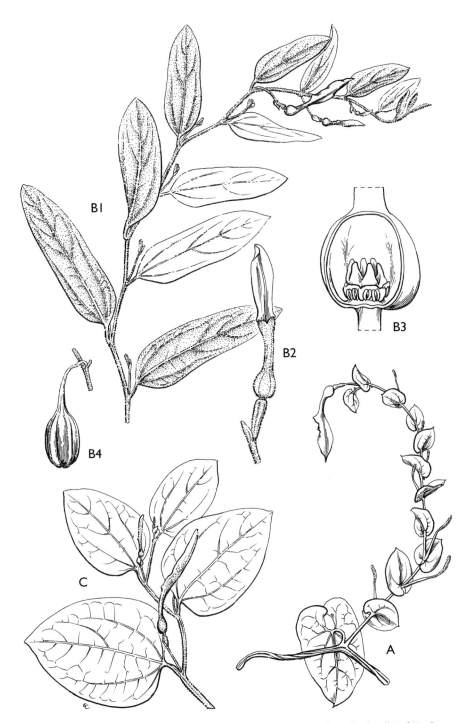

Tab. 8. A. —ARISTOLOCHIA ALBIDA, inflorescence arising from leaf axil (×2/3), from *Jacobsen* 2990. B. —ARISTOLOCHIA HOCKII. B1, apical portion of stem (×2/3), from *Fanshawe* 1580; B2, flower (×1); B3, utricle cut away to show gynostemium and anthers (×4), B2 & B3 from *Milne-Redhead* 2599; B4, fruit (×2/3), from *Fanshawe* 10099. C. — ARISTOLOCHIA HEPPII, apical portion of stem with axillary flowers (×2/3), from *Fanshawe* 8831. Drawn by Eleanor Catherine.

broad papery wing, finely verrucose on one side, the other side finely verrucose with a cuneate, winged raphe down the centre, sometimes ± covering the side.

Zambia. N: Mbala Distr., near mouth of Kalambo R. (on L. Tanganyika), c. 823 m, fl. 17.v.1936, *Burtt* 6363 (BM; BR; K). C: Mt. Makulu Res. Station, c. 19 km S of Lusaka, fl. 8.xii.1956, *Angus* 1452 (K; PRE; SRGH). S: Lusito Agric. Station, Kariba road crossing on Lusito R., fl. & fr. 9.v.1960, *Angus* 2234 (BR; K; LISC; PRE; SRGH). **Zimbabwe**. N: Hurungwe Distr., Zambezi Valley, near Rukomechi (Rekomitjie) R., fl. & fr. 21.viii.1959, *Goodier* 590 (K; LISC; PRE; SRGH). E: Mutare, Commonage, Darlington, fl. 5.ii.1949, *Chase* 1164 (BM; BR; COI; K; LISC; P; SRGH). S: Mwenezi Distr., Gonarezhou, S of Runde (Lundi) R. on bank of Nyamasikana R., fl. 28.v.1971, *Grosvenor* 558 (B; BR; K; LISC; MO; PRE; SRGH). **Malawi**. N: Lake Malawi, Likoma (Lukoma) Isl., fr. viii.1887, *Bellingham* s.n. (BM). C: Lilongwe Agric. Res. Station, fl. 25.xii.1952, *Jackson* 999 (BR; K; SRGH). S: between Muona and Shire R., 80 m, fl. 20.iii.1960, *Phipps* 2587 (K; PRE; SRGH). **Mozambique**. N: Nampula, fl. 15.i.1937, *Torre* 1193 (COI; LISC). Z: Mocuba Distr., Namagoa, 60–120 m, fl. v.1943, *Faulkner* 131 (K; PRE; SRGH). T: Boroma, near Sisitso on Zambezi R., on banks of Messenguezi (Messenguese) R., c. 305 m, fl. 9.vii.1950, *Chase* 2598 (BM; COI; K; SRGH). MS: Chemba, Chiou, C.I.C.A. Experimental Station, fl. 12.iv.1960, *Lemos & Macuácua* 81 (BM; BR; COI; K; LISC; PRE; SRGH). GI: Inhambane Prov., Govuro, andados 45 km da povoação Banamana para Machaíla, fr. 27.iii.1974, *Correia & Marques* 4231 (LMU).

Widespread in tropical Africa, from Senegal to Cameroon and the Central African Republic, and from southern Sudan to Kenya, Tanzania, Burundi, Zaire and Angola. Locally common at lower altitudes on sandy and alluvial soils, often beside rivers, in floodplain grassland, riverine and coastal thickets, and riverine and gully forests, also in mopane and mixed deciduous woodland, dry sandveld and scrub, and in cultivated and waste areas; 0–1500 m.

2. **Aristolochia hockii** De Wild. in Fedde, Repert. Spec. Nov. Regni Veg. **11**: 509 (1913); in Ann. Mus. Congo Belge, Bot., Sér 4, **2** (1): 34 (1913); Contrib. Fl. Katanga: 54 (1921). — Hauman in F.C.B. **1**: 383 (1948). —Jacobsen in Kirkia **9**: 156 (1973). —Verdcourt in F.T.E.A., Aristolochiaceae: 7 (1986). TAB. **8**, figs. B1–B4. Type from Zaire.

Aristolochia rhodesica R.E. Fr., Wiss. Ergebn. Schwed. Rhod.-Kongo-Exped. **1**, 1: 27, t. 6 (1914). Syntypes: Zambia, Mporokoso, Kunkuta R., fl. ?1911, *Fries* 1186 (UPS); Katwe, between Mporokoso and Mbala (Abercorn), fl. ?1911, *Fries* 1186a (UPS).

Aristolochia hirta Peter in Feddes Repert. Spec. Nov. Regni Veg. Beih. **40** [Peter, Fl. Deutsch-Ostafrika], part 2: 184; Descriptiones in Anhang: 17, t. 1a & 1b (1932). Type from Tanzania.

Perennial herb with annual stems from a woody rootstock. Stems climbing, trailing or suberect, up to 45 cm long, sulcate and angular (in dried state at least), sparsely to densely pubescent, often with both long and minute hairs, or tomentose. Leaves 2–14 × 0.4–3(4.5) cm, linear-lanceolate to narrowly oblong-lanceolate or ovate-lanceolate, obtuse to acute, sometimes mucronate at the apex, cuneate to rounded at the base; lamina with 3 ± parallel, longitudinal veins from the base, tomentose, or pubescent with very short hairs on the lamina and long hairs mainly concentrated along the veins particularly on the leaf lower surface, or less often the pubescence of long hairs only; venation prominent below; petioles 1.5–7 mm long, tomentose, or pubescent with a mixture of long and very short hairs. Flowers solitary, axillary, pendulous; pedicels (including ovary) 0.3–20 mm long, pubescence as for stems; inflorescence bract 2.5–12 × 1–3 mm, lanceolate to narrowly ovate, tomentose to pubescent with long and very short hairs. Perianth greenish or purplish-green in the lower part, glabrous to densely hairy; constricted part below the utricle 0.5–4 mm long; inflated part (utricle) 3–6 mm in diameter, subglobose; tube 3–15 × 1–2 mm, straight, cylindrical, dilated at apex; limb 8–35 × 3.5–11 mm, oblong-elliptic, rounded to subacute at the apex, glabrous to densely hairy outside, glabrous to glabrescent inside, reddish-purple to dark purplish-brown, buff-cream in throat. Gynostemium 2–3 mm tall, anthers 6. Fruit a pendulous 6-ribbed capsule, pubescent to glabrescent, 1.5–3 × 0.7–1.6 cm, oblong-obpyriform, apex rounded, base attenuate into stipe 7–15 mm long; the mature capsule and stipe split longitudinally into 6 segments which remain joined to the pedicel by the elements of the stipe. Seeds 3.5–7 × 3–6.5 mm, triangular with rounded corners and curled margins (in dried state at least), creamy-yellow, smooth, with thick cuneate raphe down middle of one side, sometimes ± equalling the seed in width.

Zambia. N: Chinsali, c. 1310 m, fl. & fr. 26.ix.1938, *Greenway* 5777 (BR; K; PRE). W: Mwinilunga Distr., slope E of Matonchi Farm, fr. 29.xii.1937, *Milne-Redhead* 3870 (BM; K; PRE).

C: c. 9.5 km E of Lusaka, c. 1280 m, fl. 15.x.1955, *King* 165 (K). **Zimbabwe**. N: Hurungwe (Urungwe) Res., Magunje (Mgunje), c. 1220 m, fl. 23.xi.1953, *Wild* 4245 (K; SRGH).

Also from Zaire and Tanzania. Open grassland and *Brachystegia* woodland, usually in sandy soil; 990–1310 m.

Verdcourt, in F.T.E.A., recognized two subspecies on the basis of seed morphology, subsp. *hockii* with smooth seeds and subsp. *tuberculata* Verdc. with warty seeds. Neinhuis (pers. comm.) suggests that these subspecies can be separated on leaf vein indumentum: in subsp. *hockii* the indumentum of leaf veins consists of a mixture of long and short hairs, while in subsp. *tuberculata* the leaf veins are ± densely and uniformly puberulous only and lack additional long hairs. More ripe fruiting material of this species, from throughout its range of distribution, needs to be seen.

Subsp. *tuberculata*, as delimited above, is apparently restricted to the southern highlands of Tanzania.

3. **Aristolochia heppii** Merxm. in Mitt. Bot. Staatssamml. München **6**: 197 (1953). TAB. **8**, fig. C. Type: Zimbabwe, Rusape, *Dehn* R19 (M, holotype; K; SRGH).

Perennial herb with annual stems from a woody or fleshy rootstock. Stems trailing, up to c. 40 cm long, usually sulcate and angular (in dried state at least), puberulous to shortly pilose, sometimes sparsely so. Leaves 1.5–9(10) × 1–6(8) cm, ovate to ovate-oblong, obtuse to subacute and sometimes mucronate at the apex, rounded to subtruncate at the base; lamina with 5(7) principal longitudinal veins from the base, glabrous above, usually shortly pilose on the veins below, sometimes sparsely so; veins prominent below; petioles 2–10(13) mm long, shortly pilose. Flowers solitary in leaf axils, rarely 2(3?), or in short racemose inflorescences; pedicels (including ovary) 8–15 mm long, puberulous to shortly pilose; bracts leaf-like, 4–12 × 1.5–7 mm, ovate to elliptic or lanceolate, acute at the apex, rounded to cuneate at the base, ± glabrous above, shortly pilose mainly on veins below, sometimes sparsely so. Perianth greenish in lower part, dark purple to crimson-brown in upper part, shortly pilose mainly along veins outside; limb puberulous or ± glabrous inside, pilose in throat; constricted part below the utricle 1–2 mm long; inflated part (utricle) 2.5–7 mm in diameter, subglobose; tube straight, 5–16 × 1–2 mm, cylindrical, dilated at apex; limb 13–35 × 2–8 mm, narrowly ligulate or oblong elliptic. Gynostemium 2–3 mm tall, anthers 6. Fruit a 6-ribbed capsule, 1.5–2.7 × 1–1.8 cm, oblong-pyriform, apex rounded, base attenuate into stipe 3–7 mm long, sparsely pilose mainly along ribs to glabrous; the mature capsule and stipe split longitudinally into 6 segments which remain joined to the pedicel by the elements of the stipe. Seeds 5–6.5 × 3.5–5 mm, triangular with rounded corner and curled margins (in dried state at least), creamy-yellow, smooth with thick cuneate raphe down middle of one side sometimes ± equalling the seed in width.

Zambia. W: Solwezi Distr., fl. 23.vii.1964, *Fanshawe* 8831 (K; SRGH). S: Kalomo Distr., Zimba Railway Station, c. 1250 m, fl. 1.vi.1930, *Milne-Redhead* 415 (K). **Zimbabwe**. N: Gokwe Distr., Sengwa Res. Station, Big Air Strip, fr. 21.vi.1983, *Mahlangu* 753 (K; SRGH). C: Chegutu Distr., Poole Farm, fl. & fr. 3.iv.1946, *Wild* 989 (K; SRGH). E: Mutare Distr., Zimunya Res., c. 915 m, fl. & fr. 20.iii.1955, *Chase* 5525 (BM; K; SRGH). S: Masvingo Distr., Mushandike Nat. Park, Mushandike R., fr. vii.1971, *P.J. Wright* T290 (K; SRGH).

Also from Zaire. Open grassland and sandveld, also in miombo and *Brachystegia* woodlands, usually in sandy soil; 915–1250 m.

Related to *A. hockii* but differing mainly in its broader, ± truncate leaves and predominantly creeping habit.

4. **Aristolochia elegans** Mast. in Gard. Chron. **24**: 301, fig 64 (1885). —Hooker f. in Bot. Mag. **112**, t. 6909 (1886). —Baker & C.H. Wright in F.T.A. **6**, 1: 138 (1909). —Cufodontis, Enum. Pl. Aethiop. Sperm.: 36 (1953); Keay in F.W.T.A., ed. 2, **1**: 81 (1954). —Jex-Blake, Gard. E. Afr., ed. 4: 131 (1957). —Verdcourt in F.T.E.A., Aristolochiaceae: 10 (1986). Syntypes from Brazil and cult. in England.

Glabrous perennial climbers, woody at the base. Stems slender 2–3(6) m long, sulcate (in dried state at least), flowering branches pendulous. Leaves 1.5–11 × 20–11 cm, broadly ovate-reniform, apex usually broadly rounded, rarely subacute, base strongly cordate with a broad sinus, slightly cuneate into petiole; lamina 5–7-nerved from the base, mid-green on upper surface, paler grey-green below; petioles 7–60 mm long; pseudostipules leaf-like, 0.6–2.5 × 1–2.5 cm, rotund to reniform, cordate. Flowers solitary, axillary; pedicels (including ovary) 5–12 cm long.

Perianth: inflated part (utricle) 20–35 × 8–20 mm, ellipsoid, sometimes hairy inside, greenish-cream or yellow; tube 8–30 × 1.5–9 mm, cylindrical, dilated at apex, sharply reflexed back over basal inflated part, hairy inside, greenish-cream; limb excentric about the throat, 30–110 × 26–80 mm, ovate to rotund, 2-lobed at the base, rounded, sometimes mucronate or emarginate at the apex, creamy-white, with crimson reticulate markings externally, with dense feathery crimson-brown mottling internally and dark crimson blotch around the yellow velvety throat. Gynostemium 5–7 mm tall, anthers 6. Fruit 3–5 × 1–1.5(2) cm, oblong-cylindrical, 6-ribbed, rostrate, dehiscing septicidally from base upwards into 6 valves that remain attached at apex forming a "hanging basket". Seeds 4–6.5 × 3–5 mm, obovate, flat, warty on one side, less warty but with a median line on the other, the apical indentation and narrow marginal wing ± smooth on both sides.

Zimbabwe. C: Chegutu Distr., Chegutu Junior School grounds, fr. 21.vi.1968, *Rushworth* 1192 (SRGH). **Malawi**. S: Zomba Distr., 11 km E of Zomba, Mingoli Farm Estate by Likangala R., 730 m, fr. 29.xi.1977, *Brummitt, Seyani & Dudley* 15216 (K; SRGH). **Mozambique**. Z: Mocuba, Namagoa, 60–120 m, fl. & fr. v.1945, *Faulkner* Pretoria No. 212 (K; SRGH).

Native to Brazil, Paraguay and N Argentina. Widely cultivated and naturalized in tropical and temperate regions of the world. In evergreen riverine forest and forest margins, usually in disturbed areas.

Verdcourt (loc. cit.) pointed out that the earlier name, *A. littoralis* Parodi in Anal. Soc. Cient. Argent. **5**: 155 (1878), has been suggested as the correct name for this species by various authors including: Hoehne in Fl. Brasilica **15**, 2: 47, t. 10 (1942); Hauman in Ann. Mus. Nac. Hist. Nat. Buenos Aires **32**: 328, t. 2/9 (1923); Hauman & Irigoye in Ann. Mus. Nac. Hist. Nat. Buenos Aires **32**: 61 (1923); Pfeifer in Ann. Missouri Bot. Garden **53**: 160 (1966). I have accepted his preference for the totally unambiguous name *A. elegans*.

142. PIPERACEAE

By M.A. Diniz

Annual or perennial, often succulent herbs with erect or creeping stems, sometimes epiphytic, or erect sometimes scrambling or climbing subshrubs or shrubs, or lianes, occasionally slender trees (not in the Flora Zambesiaca area), without latex, frequently aromatic; plants dioecious, monoecious or with hermaphrodite flowers. Stems with distinct vascular bundles, sometimes scattered as in Monocotyledons, often with jointed and swollen nodes, and often ± zigzag. Leaves usually alternate, less often opposite or verticillate, simple, entire, petiolate, thin to very succulent, often glandular-pellucid, pinnately or palmately nerved. Stipules adnate to the petiole, or absent. Inflorescences terminal, axillary or leaf-opposed with flowers in simple, dense and ± fleshy spikes, more rarely in racemes or spikes grouped in umbels. Flowers minute, mostly hermaphrodite, sometimes unisexual or polygamous, each flower subtended by a minute bract; perianth absent. Stamens 2–6 (1–10); filaments thick, usually free; anthers 2-thecous, erect, distinct or confluent, basifixed or dorsifixed, dehiscing longitudinally. Ovary superior, 1-locular, 1-ovulate; ovule basal, orthotropous, erect; styles 1–5 or absent; stigmas 1–5 capitate, linear or penicillate, central or excentric. Fruit an indehiscent drupe (considered a berry by some authors) small, free, sessile or pedunculate with succulent or thin dry pericarps. Seed ± globose, with little endosperm but copious perisperm; embryo very small.

A family of c. 7 genera and more than 1000 species occurring in the tropical regions, mainly in S America and S Asia. Only 2 genera occur in Africa.

A number of species of *Piper* are important condiments: *P. betle* L., *P. cubeba* L.f., *P. guineense* Thonn. ex Schumach., *P. longum* L., *P. methysticum* Forst.f., *P. miniatum* Blume, *P. nigrum* L. and *P. sylvaticum* Roxb.

Shrubs, climbers or lianes, not succulent; leaves alternate; petioles with adnate sheathing stipules; stamens 2–6; anther-thecae usually distinct; stigmas 2–5 · · · · · · · · · · · · 1. **Piper**
Herbs, usually epiphytic, often succulent; leaves alternate, opposite or verticillate; stipules absent; stamens 2; anther-thecae confluent; stigmas 1 · · · · · · · · · · · · · · · 2. **Peperomia**

1. PIPER L.

Piper L., Sp. Pl. : 28 (1753); Gen. Pl., ed. 5: 18 (1754). —Bentham & Hooker f., Gen. Pl. **3**: 129 (1880). —Tebbs in Kubitzki et al., Fam. Gen. Vasc. Pl. **2**: 518 (1993).

Erect, scandent or climbing shrubs, or sometimes lianes. Stems often hollow, jointed at the nodes. Leaves alternate, sometimes asymmetrical at the base, membranous or coriaceous. Stipules adnate to the petiole, soon caducous. Inflorescences axillary or leaf-opposed, pedunculate, with flowers in dense cylindrical spikes, rarely racemose. Flowers hermaphrodite or unisexual; plants monoecious or dioecious. Stamens 2–4(6) in the Flora Zambesiaca area; filaments short; anthers basifixed with the thecae usually distinct. Ovary ± ovoid to subglobose; stigmas 2–4(5), erect or recurved, sessile or on a short style. Fruit a ± globose 1-seeded drupe, sessile, rarely pedunculate. Seed subglobose; testa thin; endosperm and perisperm hard.

A large genus of about 1000 species confined almost exclusively to tropical forest regions of both hemispheres, especially abundant in tropical America. Very few occur naturally in Africa, and these are common and widespread.

1. Spikes 2–8, umbellately arranged on leafless axillary branchlets; leaves large, subcircular to reniform, deeply cordate; flowers bisexual · 1. *umbellatum*
– Spikes solitary, leaf-opposed; leaves smaller, ovate, elliptic or oblong-lanceolate, cuneate or rounded at the base or ± cordate but not deeply so; flowers bisexual or unisexual by abortion · 2
2. Leaves palmately 3–7-nerved from the base; stems without adventitious roots; stigmas 2; fruits sessile · 2. *capense*
– Leaves pinnately nerved; stems with adventitious roots; stigmas 3; fruits pedunculate · 3. *guineense*

1. **Piper umbellatum** L., Sp. Pl. : 30 (1753). —C. De Candolle in De Candolle, Prodr. **16**, 1: 332 (1869). —Hiern, Cat. Afr. Pl. Welw. **1**: 911 (1900). —Baker & C.H. Wright in F.T.A. **6**, 1: 144 (1909). —Hutchinson & Dalziel in F.W.T.A. **1**: 80, fig. 79 (1927). —Balle in Bull. Jard. Bot. État **16**: 36 (1942). —Drummond in Kirkia **10**: 233 (1975). —Beentje, Kenya Trees Shrubs Lianas: 66 (1994). —Verdcourt in F.T.E.A., Piperaceae: 8 (1996). TAB. **9**. Type from the Dominican Republic.

Piper subpeltatum Willd., Sp. Pl. **1**: 166 (1797). —Engler, Pflanzenw. Ost-Afrikas **C**: 159 (1895). Syntypes from Indonesia.

A shrub up to 3(4) m high, with numerous stems arising from a woody rootstock, becoming scrambling. Stems subsucculent, thick, slightly striate, glabrous, with adventitious roots. Petiole up to 30 cm long, dilated and amplexicaul at the base, grooved above, terete beneath, glabrous, glandular-punctate. Leaf lamina (5)6–36(40) × (4.5)6–32(42) cm, subcircular to reniform, ± shortly acuminate at the apex, deeply cordate at the base, membranous, with a mint-like fragrance, discolorous, dark green and slightly shining above, lighter green below and ± glandular-punctate, sparsely to densely pubescent on the nerves above and beneath and also on the conspicuously reticulate venation beneath, ± ciliate at the margins; nerves 11–13(15), palmate, impressed above and prominent below, reticulation conspicuous below. Inflorescences glabrous, consisting of 2–8 spikes umbellately arranged on reduced leafless axillary shoots 4–12 cm long; umbel bracts up to 8 mm long, deciduous; spikes (4)5–10(12) cm long and c. 0.3 cm in diameter, bright greenish-white, glabrous, with slender peduncles 0.4–1.7 cm long. Flowers bisexual; floral bracts 0.5 mm across triangular to ± subcircular, distinctly white fimbriate at the margins, petiolulate; stamens 2; stigmas 3. Fruit c. 0.6–0.8 × 0.4–0.5 mm, obpyramidal, trigonous, brownish when ripe, glandular-punctate.

Zimbabwe. E: Chipinge Distr., Chirinda Forest, c. 1160 m, fl. & fr. iii.1962, *Goldsmith* 108/62 (COI; K; LISC; SRGH). **Malawi**. S: Mt. Mulanje south slopes, Esperanza Estate, 750 m, 17.ix.1983, *Dowsett-Lemaire* 998 (K). **Mozambique**. Z: Lugela, 5.1 km from Limbue towards Murua R., fl. 25.v.1949, *Barbosa & Carvalho* 2873 (K; LMA). MS: Manica Prov., Chimoio, Serra do Garuso, fl. & fr. 24.iii.1948, *Mendonça* 3856 (LISC).

Pantropical, including west and central Africa from Guinea-Bissau to Zaire and Angola, São Tomé and Bioko, and in S Sudan, Uganda, Kenya and Tanzania, and the Seychelles, Madagascar and the Mascarene Islands. Low to medium altitude evergreen rainforest understorey and in moist shady riverine forest; 400–1200 m.

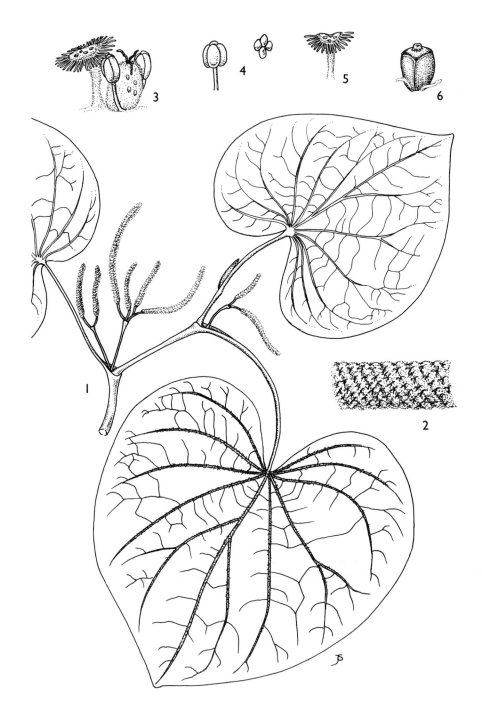

Tab. 9. PIPER UMBELLATUM. 1, twig with inflorescences (×²/₃), from *Barbosa & Carvalho* 2873; 2, portion of inflorescence (×6); 3, flower with subtending bract (×32); 4, stamen, viewed from side and above (×32); 5, floral bract (×32); 6, fruit with remains of stamens (×24), 2–6 from *Cholocholo, Nurmi & Steiner* 31. Drawn by Judi Stone.

2. **Piper capense** L.f., Suppl. Pl.: 90 (1782). —Thunberg, Fl. Cap., ed. **2**: 443 (1823). —C. De Candolle in De Candolle, Prodr. **16**, 1: 339 (1869); in Bot. Jahrb. Syst. **19**: 224 (1894). —Engler, Pflanzenw. Ost-Afrikas **C**: 159 (1895). —Burkill in Johnston, Brit. Cent. Africa [Bot. Appendices]: 267 (1897). —Baker & C.H. Wright in F.T.A. **6**, 1: 146 (1909). —Mildbraed, Wiss. Ergebn. Deutsch. Zentr.-Afr.-Exped., Bot., part 3: 177 (1911). —S. Moore in J. Linn. Soc., Bot. **40**: 183 (1911). —C.H. Wright in F.C. **5**, 1: 488 (1912). —Engler, Pflanzenw. Afrikas (Veg. Erde 9) **3**, 1: 3, fig. 2 E–G (1915). —Eyles in Trans. Roy. Soc. South Africa **5**: 337 (1916). —Hutchinson & Dalziel in F.W.T.A. **1**: 80 (1927). —Balle in Bull. Jard. Bot. État **16**: 370 (1942). —Brenan in Mem. New York Bot. Gard. **9**: 58 (1954). —Topham, Check List For. Trees Shrubs Nyasaland Prot. ed. 2: 80 (1958). —Goodier & Phipps in Kirkia **1**: 54 (1961). —F. White, F.F.N.R.: 19, fig. 3 (1962). —Drummond in Kirkia **10**: 233 (1975). —Beentje, Kenya Trees Shrubs Lianas: 66 (1994). —Verdcourt in F.T.E.A., Piperaceae: 5 (1996). Type from South Africa.

Piper volkensii C. DC. in Bot. Jahrb. Syst. **19**: 225 (1894). Type from Tanzania.

Piper sacleuxii C. DC. in Not. Syst., Paris **2**: 52 (1911). Type from Tanzania.

Piper trichopodum C. DC. in Not. Syst., Paris **2**: 51 (1911). Type from Tanzania.

Piper bequaertii De Wild. in Rev. Zool. Bot. Africaines **8**, Suppl. Bot.: 10 (1920). Syntypes from Zaire.

Piper humblotii C. DC. in Candollea **1**: 178 (1923). Type not cited, but specimen from Comoro Island.

An aromatic evergreen shrub or subshrub from a tuberous rootstock, ± erect to c. 3 m high or with scandent or trailing stems 3–4(5) m long. Stems terete, weak, greenish, with glabrous to ± villous swollen nodes; nodes up to 3 cm wide. Petiole 0.5–5(8.3) cm long, grooved above near the base, terete below, glabrous or ± villous; stipules 1–1.6 cm long, lanceolate, membranous, deciduous. Leaves 5–15.5(18.5) × 2.5–15(17.5) cm, ovate, rarely elliptic or elliptic-ovate, acuminate at the apex, ± asymmetric and cuneate at the base to rounded or shortly cordate at the base, membranous, discolorous, dark green on upper surface, light green or greyish to glaucous below, glabrous on both surfaces with the nerves ± densely pilose-pubescent towards the base on the lower surface, or the lower surface pilose-pubescent all over, palmately (3)5–11-nerved with the 3 median nerves reaching the apex, impressed above, prominent below. Inflorescences consisting of solitary leaf-opposed spikes, terminal, or lateral when overtopped by sympodial growth, creamy-white; peduncles (1)1.5–5.2 cm long, glabrous or ± pilose-pubescent; rhachis (0.7)2.5–8.5(10) cm long, glabrous. Flowers: bisexual and male in separate spikes, or both on one spike with male flowers towards the base and hermaphrodite flowers towards the apex; bracts 1–1.5 mm in diameter, subcircular, glabrous above, densely pubescent below and on the petiolule; stamens 2–3; ovary ovoid; style short; stigmas 2, recurved. Fruit sessile 2–4 mm long, spheric-ovoid, obtuse.

This species exhibits much variation in spike length and the presence or absence of indumentum on the leaves. However, while these characters on their own are of little taxonomic value, in combination they delimit taxa at varietal level.

Leaves mostly glabrous on the lower surface, or pubescent on the nerves; upper shoots and petioles glabrous or slightly pubescent. Spikes short to long, up to 10 cm long · var. *capense*
Leaves ± densely pubescent on the lower surface; upper shoots and petioles ± densely pubescent. Spikes short, 0.7–2 cm long · var. *brachyrhachis*

Var. **capense** —Verdcourt in F.T.E.A., Piperaceae: 5 (1996).

Zambia. N: Mbala Distr., Sunzu Mt., 1800 m, fr. 23.iv.1962, *Richards* 16376 (K; SRGH). W: Mwinilunga Distr., Matonchi R., near Matonchi Farm, fl. & fr. 15.x.1937, *Milne-Redhead* 2792 (BM; K). E: Nyika Plateau, 8.8 km SW of Rest House, 2150 m, fl. 25.x.1958, *Robson* 348 (K; SRGH). **Zimbabwe**. E: Chimanimani Distr., Chirinda Forest, 1050 m, fr. 14.v.1953, *Chase* 4968 (BM; SRGH). **Malawi**. N: Rumphi Distr., Livingstonia Escarpment, Nyamkowa, in Nyika Nat. Park, 2000 m, fl. 23.ii.1978, *Pawek* 13864 (K; MA; MO). C: Ntchisi Mt., 1450 m, fl. & fr. 19.ii.1959, *Robson* 1658 (K; LISC; SRGH). S: Mulanje Distr., above Chisambo Tea Estates, valley leading towards Manene Peak, c. 1400 m, fl. 3.ix.1970, *Müller* 1528 (K; SRGH). **Mozambique**. N: Nampula Prov., Ribáuè Distr., Serra Ribáuè (Mepáluè), 1450 m, fl. 28.i.1964, *Torre & Paiva* 10295 (COI; LISC; LMU; PRE; WAG). Z: Milange, Serra Milange, 1300 m, fl. & fr. 26.ii.1943, *Torre* 4836 (BR; LISC; LMA; SRGH). T: Angónia Distr., Serra Dómuè, fr. 19.vii.1979, *Macuácua*

& Stefanesco 936 (LMA). MS: Manica Prov., Serra Zuira, Plateau Tsetserra (Tsetsera), 7 km from cow-house, road to Chimoio (Vila Pery), 1840 m, fl. 12.xi.1965, *Torre & Pereira* 12920 (BR; EA; LISC; LMA; M).

Widespread in Africa from Sierra Leone to Cameroon, São Tomé and Bioko to Zaire and Angola, Ethiopia and Sudan to Rwanda, Burundi, Uganda, Kenya, Tanzania and South Africa, also in Comoro Islands and Madagascar. Evergreen rainforest understorey, swamp forest (mushitu) and in moist shady riverine forest; 650–2250 m.

Var. **brachyrhachis** (C.H. Wright) Verdc. in F.T.E.A., Piperaceae: 7 (1996).
    Piper brachyrhachis C.H. Wright in F.T.A. **6**, 1: 147 (1909). —Brenan in Mem. New York Bot. Gard. **9**, 1: 58 (1954). Syntypes: Malawi, between Mpata and the commencement of the Tanganyika Plateau, 610–915 m, vii.1896, *Whyte* s.n. (K); Nyika Plateau, 1210–2135 m, vi.1896, *Whyte* 199 (K).

Zambia. N: 40 miles south of Ishiba Ngandu, *Angus* 854 (FHO; K); Lunzuwa south of Mbala, *Angus* 1996 (FHO; K); Mafingas, *Chisumpa* 68 (K; NDO), half way between Kasama and Mbala, *Hooper & Townsend* 1891 (K); Lupita–Makutu, *Lawton* 1273 (K); Chisau R. gorge, *Richards* 11788 (K); Lake Chila near Mbala, *St. Clair-Thomson* 1092 (K). W: Chingola, *Fanshawe* 2420 (K); Solwezi, *Milne Redhead* 449 (K). C: Mkushi, *Fanshawe* 2082 (K); Serenje, *Simon & Williamson* 1622 (K; SRGH). **Malawi.** N: Viphya, *Pawek* 3998 (K; MAL); Jembeya For. Res., *Thompson & Rawlins* 5527 (K). C: Chongoni Mt., *Iwarsson & Ryding* 887 (K); Dedza, *Adlard* 417 (K); Robson 1418 (K); *Salubeni* 1066 (K; SRGH); Ntchisi Mt., *Brass* 16937 (K); *Salubeni* 603 (K; SRGH).

Also in SW Tanzania. From evergreen forest, moist riverine forest shade, mushitu (swamp forest) and high rainfall miombo woodland.

Piper brachyrhachis var. *glabra* Balle from Zaire, with young leaves more lanceolate and the 3 inner nerves meeting above the base, is probably a third variety.

3. **Piper guineense** Thonn. ex Schumach., Beskr. Guin. Pl.: 19 (1827). —C. De Candolle in De Candolle, Prodr. **16**, 1: 343 (1869). —Hiern, Cat. Afr. Pl. Welw. **1**: 912 (1900). —Baker & C.H. Wright in F.T.A. **6**, 1: 145 (1909). —Engler in Mildbraed, Wiss. Ergebn. Deutsch. Zentr.-Afr.-Exped., Bot., part 3: 177 (1911); Pflanzenw. Afrikas (Veg. Erde 9) **3**, 1: 3, fig. 2 A–D (1915). —Thonner, Flow. Pl. Afr.: 161, fig. 27 (1915). —Hutchinson & Dalziel in F.W.T.A. **1**: 80 (1927). —Balle in Bull. Jard. Bot. État **16**: 375, t. 16 (1942). —Keay in F.W.T.A., ed. 2, **1**: 84 (1954). —F. White, F.F.N.R.: 19 (1962). —Beentje, Kenya Trees Shrubs Lianas: 66 (1994). —Verdcourt in F.T.E.A., Piperaceae: 4 (1996). Type from Ghana.

A climbing shrub, or liane to 9(23) m tall, climbing into trees by means of adventitious roots; stems glabrous or ± pubescent, older stems with corky ridges, terminal branches erect or pendent; adventitious roots arising near the nodes of the main stems, absent from secondary branches. Leaves very variable in shape according to position on the plant; petiole 0.5–4 cm long, grooved above, terete beneath, longer in the leaves of primary branches; lamina 4.5–14(16.5) × (2)2.5–7(12) cm, ovate to elliptic, rarely subcircular, long-acuminate at the apex, symmetrical or cordate at the base in the lower leaves and rounded in the upper ones of the primary branches, rounded cordate or cuneate at the base of leaves on secondary branches, papyraceous to coriaceous, dark dull green on upper surface, pale blue-green beneath, glabrous on both surfaces or sometimes pubescent on the nerves beneath; nervation pinnate, lateral nerves in about 2–4 pairs, conspicuous and impressed above, prominent beneath. Inflorescence of terminal or lateral leaf-opposed solitary spikes; peduncles 0.5–1(2.5) cm long, usually glabrous; rhachis 2.5–3.0 cm long, increasing to 10 cm long in fruit, ± pubescent; bracts c. 1 mm in diameter, ± circular or obovate, peltate, ciliate at the margin, pubescent beneath; stamens 2; ovary ovoid; stigmas 3(5), sessile. Fruit (3)4–5(6) mm in diameter, subglobose to obovoid, thinly fleshy, orange-greenish becoming scarlet, with peduncle 4–10 mm long when ripe.

Zambia. W: Mwinilunga Distr., Matonchi R., c. 1.6 km south of Matonchi Farm, fr. 7.ii.1938, *Milne-Redhead* 4487 (K).

Widespread in tropical Africa from Mali and Guinea Bissau to Angola and Zaire and to Uganda, Kenya and Tanzania, also in São Tomé and Bioko. In evergreen riverine forest and ± extensive in swamp forest (mushitu), also in muteshi woodland understorey; up to 1650 m.

142. PIPERACEAE 29

2. PEPEROMIA Ruiz & Pav.

Peperomia Ruiz & Pav., Fl. Peruv. Chilen. Prodr. : 8, t. 2 (1794). —Düll in Bot. Jahrb. Syst. **93**: 56–129 (1973).

Annual or perennial often succulent herbs, terrestrial epiphytic or growing on rocks; stems erect or ascending, sometimes climbing. Leaves alternate, opposite or verticillate, simple, entire, sometimes peltate, thin to ± fleshy, often glandular-punctate, glabrous or hairy, stipules absent. Inflorescences of solitary or crowded spikes, terminal, leaf-opposed or axillary. Flowers hermaphrodite, numerous, sessile, sometimes sunk in depressions in the rhachis, usually white cream or green. Stamens 2, borne at the base of the ovary, soon deciduous; filaments short, very thin; anthers transversely oblong or subglobose, dorsifixed. Ovary globose or ellipsoid, glabrous; stigma 1(2), sessile or style developed, simple, sometimes sunken, smooth or penicillate. Fruit a sessile or stipitate drupe; pericarp thin. Seed 1, testa membranous or coriaceous; endosperm farinaceous.

A large pantropical genus with c. 1000 species, the majority in America, others in Asia and about 17 in tropical Africa. Several species are cultivated as house plants.

1. Leaves all 3–4-verticillate; fruit with a glanduliferous pseudocupule at the base · 1. *tetraphylla*
– Leaves opposite or alternate, sometimes the upper ones 3–4-verticillate; fruit without pseudocupule · 2
2. Leaves opposite, (rarely alternate at the base), sometimes 3–4-verticillate at base of the spikes, pubescent on both surfaces; stems pubescent · · · · · · · · · · · · · · · · · · · 2. *blanda*
– Leaves alternate (rarely opposite at the base), glabrous or if pubescent then not on both surfaces; stems glabrous · 3
3. Leaves circular to cordate or obovate-cordate to almost reniform, up to 3 cm long, very thin and translucent, glabrous; fruit ovoid-spheric with warty-dentate longitudinal ridges; plants erect, annual or ephemeral · 3. *pellucida*
– Leaves if circular then up to 1 cm long, not translucent, sparsely hairy sometimes only ciliate on the margins; fruit not as above; plants trailing or creeping, annual or perennial · · · · 4
4. Plants creeping, with only the spikes erect; rhachis up to 1 cm long, shorter than the peduncle, which may reach 2 cm long; leaves circular to elliptic, up to 1 cm long · · · · · · 4. *rotundifolia*
– Plants trailing and creeping with erect branches; rhachis mostly more than 1 cm long, always longer than the peduncle; leaves variously shaped, sometimes the basal ones circular · · · 5
5. Leaves small, up to 2.5(3) cm long, pinnately veined · · · · · · · · · · · · · · · · · · 5. *retusa*
– Leaves mostly more than 2.5 cm long, (except the lower ones), palmately 3–5(7)-nerved · · 6
6. Leaves spathulate to obovate-oblong, apex rounded to obtusely acuminate and sometimes emarginate · 6. *goetzeana*
– Leaves not as above · 7
7. Leaves broadly ovate to broadly lanceolate, papery, ± acuminate, acute or sometimes obtuse at the apex, never emarginate; petiole (4)6–15(20) mm long, slender · · · · · · · · 7. *molleri*
– Leaves rhomboid-elliptic, elliptic, sometimes ovate, chartaceous, obtuse and usually emarginate at the apex; petiole (2)3–5(10) mm long, thick · · · · · · · · · · · · 8. *abyssinica*

1. **Peperomia tetraphylla** (G. Forst.) Hook. & Arn., Bot. Beechey Voy.: 97 (1832). —Düll in Bot. Jahrb. Syst. **93**: 72 et 74, fig. 2 (1973). —Verdcourt in F.T.E.A., Piperaceae: 12 (1996). Type from Society Islands.
    Piper reflexum L.f., Suppl. Pl.: 91 (1782). Type from South Africa.
    Piper tetraphyllum G. Forst., Fl. Ins. Austr. Prodr.: 5 (1786).
    Peperomia reflexa (L.f.) A. Dietr., Sp. Pl., ed. 6, **1**: 180 (1831). —Engler, Pflanzenw. Ost-Afrikas **C**: 159 (1895). —Baker & C.H. Wright in F.T.A. **6**, 1: 155 (1909). —S. Moore in J. Linn. Soc., Bot. **40**: 183 (1911). —C.H. Wright in F.C. **5**: 490 (1912). —Engler, Pflanzenw. Afrikas (Veg. Erde 9) **3**, 1: 5, fig. 3A (1915). —Eyles in Trans. Roy. Soc. South Africa **5**: 338 (1916). —Weimarck in Bot. Notis. **1936**: 1 (1936). —Balle in Bull. Jard. Bot. État **16**: 383 (1942). —Brenan in Mem. New York Bot. Gard. **9**: 60 (1954). —Keay, F.W.T.A., ed. 2, **1**: 83 (1954). —Goodier & Phipps in Kirkia **1**: 54 (1961) nom. illegit., non Kunth (1816).

Perennial, usually epiphytic herbs with tufted creeping stems 6–20 cm long, sometimes forming mats on tree branches, the stolons giving rise to more erect flowering shoots. Stems slender, simple or dichotomously branched, rooting at the lower nodes, 4-angled; lower internodes up to 5 cm long, very short towards the apex. Petiole subsessile or up to 15–30 mm long. Leaves (3)4-whorled, (4)6–15(18)

× (3)4–11(12.5) mm, elliptic, rhomboid, ovate, obovate-elliptic or almost rounded, broadly obtuse to rounded at the apex, rounded or broadly cuneate at the base, glabrous on upper surface, glabrous to subglabrous or puberulous beneath, thick, coriaceous, weakly shining, drying ± greyish-glaucous, longitudinally and ± obscurely 3-nerved. Inflorescence of solitary, terminal or axillary, green spikes; rhachis (8)12–30(40) mm long, shortly hairy; peduncle (6)10–18(25) mm long, glabrous. Ovary ovoid, sunk in alveoli on the rhachis; stigma terminal. Fruit c. 1 mm long, subcylindrical, with a pseudocupule at the base, covered with viscous glands.

Zambia. B: Kabompo R., fl. 23.x.1966, *Leach & Williamson* 13446 (K; LISC). N: Mpika Distr., Muchinga Escarpment, 45 km NE of Mpika, fr. 29.xi.1952, *White* 3794 (K). W: Kabompo R. crossing, 95 km E of Mwinilunga, fr. 15.iv.1960, *E.A. Robinson* 3552 (K; SRGH). C: Serenje Distr., Kundalila Falls, 12.8 km SE of Kanona, 1400 m, fl. 17.xii.1967, *Simon & Williamson* 1423 (K; SRGH). E: Nyika, fl. & fr. 26.vi.1966, *Fanshawe* 9771 (K; SRGH). **Zimbabwe**. C: Wedza Distr., Wedza Mt., fl. & fr. 21.v.1968, *Mavi* 728 (K; LISC; SRGH). E: Mutare Distr., Engwa, 1830 m, fl. 3.ii.1955, *Exell, Mendonça & Wild* 169 (BM; LISC). S: Bikita Distr., Turgwe Gorge, below Dafana confluence, 925 m, fl. & fr. 8.v.1969, *Biegel* 3064 (K; SRGH). **Malawi**. N: Nkhata Bay Distr., 8 km E of Mzuzu at Roseveare's Cottage, 1220 m, fl. & fr. 7.vii.1975, *Pawek* 9679 (MO; SRGH; UC). C: Dedza Distr., Chiwau Hill, fl. & fr. 17.vii.1969, *Salubeni* 1372 (K; SRGH). S: Malosa St., 1220 m, st. xi/xii.1896, *Whyte* s.n. (K). **Mozambique**. N: Niassa Prov., Lichinga (Vila Cabral), Serra Massangulo, 1450 m, fl. 25.ii.1964, *Torre & Paiva* 10805 (BR; LISC; LMA). Z: Gurué, Serra Murece, 1100 m, fl. & fr. 23.vii.1979, *Schäfer* 6783 (BM; K; LMU). T: Zóbuè, Zóbuè Mt., fl. & fr. 3.x.1942, *Mendonça* 608 (LISC; LMA). MS: Manica Prov., Serra Zuira, Plateau Tsetserra, 1800 m, fl. 6.ii.1965, *Torre & Pereira* 12720 (BR; LISC; LMU).

Pantropical, and widespread in Africa from Guinea to Cameroon and São Tomé, Zaire to S Sudan, Ethiopia, Somalia, Uganda, Kenya, Tanzania and South Africa. In evergreen rain forest, gully forest and moist evergreen riverine forest, also in mist forest and woodland, epiphytic on living or fallen trees and on rocks; 900–2200 m.

2. **Peperomia blanda** (Jacq.) Kunth in Humboldt Bonpland & Kunth, Nov. Gen. Sp. **1**: 67 (1816). —Düll in Bot. Jahrb. Syst. **93**: 109 (1973). —Verdcourt in F.T.E.A., Piperaceae: 18 (1996). Type from Venezuela.

Piper blandum Jacq., Coll. Bot. **3**: 211 (1791).

An erect succulent rhizomatous herb, up to 50 cm tall (including the inflorescence). Stems decumbent, rooting at the lower nodes, simple or few-branched, densely pubescent or ± glabrous. Leaves opposite, the lowermost smaller and sometimes alternate, the upper ones at the base of inflorescence spikes 3–4-whorled, thinly succulent when fresh, drying rather membranous; petiole (4)7–15(23) mm long, pubescent or glabrescent; lamina (1.5)2.0–5.0(6.3) × (1.0)1.3–2.5(3.0) cm, broadly elliptic, ovate or obovate-oblong, rounded to obtuse or acute at the apex, cuneate at the base, pubescent to subglabrous on both surfaces, ciliate on the margin, sometimes obscurely pellucid-punctate more conspicuous on the lower surface, 3–5-palmatinerved. Inflorescences spicate; spikes terminal or axillary, greenish, 2–5 clustered, rarely solitary; rhachis (4.5)5.5–9.3(11) cm long, glabrous; peduncle (0.8)1.2–1.7(2.3) cm long, pubescent or subglabrous; flowers lax. Ovary ovoid, glabrous; stigma terminal, oblique. Fruit 0.7–0.9 mm long, globose-ovoid, densely covered with viscous shortly cylindrical papillae. Pseudopedicels present or absent.

Var. **leptostachya** (Hook. & Arn.) Düll in Bot. Jahrb. Syst. **93**: 110 et 80, fig. 16 (1973). TAB. **10**. Type from Hawaii.

Peperomia leptostachya Hook. & Arn., Bot. Beechey Voy.: 70 (1832).

Peperomia arabica Decne. in Miquel, Syst. Piper. **1**: 121 (1843). —C. De Candolle in De Candolle, Prodr. **16**, 1: 442 (1869). —Engler, Pflanzenw. Ost-Afrikas C: 159 (1895); in Mildbraed, Wiss. Ergeb. Deutsch. Zentr.-Afr.-Exp., Bot., part 3: 177 (1911). —C.H. Wright in F.C. **5**: 490 (1912). —Engler, Pflanzenw. Afrikas (Veg. Erde 9) **3**, 1: 5 (1915). —Balle in Bull. Jard. Bot. État **16**: 386 (1942). Syntypes from Yemen and South Africa.

Peperomia arabica var. *parvifolia* C. DC. in Bot. Jahrb. Syst. **19**: 230 (1894). —Baker & C.H. Wright in F.T.A. **6**, 1: 154 (1909). —Fries, Wiss. Ergebn. Schwed. Rhod.-Kongo-Exp. **1**: 11 (1914). Type from Tanzania.

Peperomia brachytrichoides Engl., Bot. Jahrb. Syst. **45**: 277 (1910). —Fries, Wiss. Ergebn. Schwed. Rhod.-Kongo-Exped. **1**, 1: 11 (1914). —Eyles in Trans. Roy. Soc. South Africa **5**: 337 (1916). Type: Zimbabwe, Victoria Falls/Sambesi, Schluchtwald, Palm-Kloof, c. 900 m, st. ix.1905, *Engler* 2917a (B, holotype).

Peperomia kyimbilana C. DC., in Bot. Jahrb. Syst. **57**: 19 (1920). Type from Tanzania.

Tab. 10. PEPEROMIA BLANDA var. LEPTOSTACHYA. 1, habit (×⅔); 2, detail of leaf upper surface and margin (×20), 1 & 2 from *Schäfer* 6737; 3, portion of inflorescence showing floral bracts and anthers (×24); 4, floral bract (×24); 5, pistil and stamens (×24); 6, portion of infructescence (×24); 7, fruit with bract (×24), 3–7 from *Hooper & Townsend* 17. Drawn by Judi Stone.

Peperomia bequaertii De Wild. in Rev. Zool. Bot. Africaines **8**, Suppl. Bot.: 6 (1920). Type from Zaire.

Peperomia spp. [*W.H. Johnson* 181] sensu Eyles in Trans. Roy. Soc. South Africa **5**: 338 (1916).

Stems and leaves ± puberulous; leaves less than 5 cm long; mature fruits without pseudopedicels.

Zambia. S: Livingstone Distr., Victoria Falls, Palm Grove, fl. & fr. iii.l918, *Eyles* 1292 (BM; K; SRGH). **Zimbabwe**. E: Chipinge Distr., Muhizu Kraal, c. 8 km N of Tanganda Tea Estate, fl. & fr. 6.viii.1973, *Mavi* 1476 (K; SRGH). S: Chibi Distr., Nyoni (Inyoni) Range, top of kloof, c. 750 m, fl. & fr. 9.v.1970, *Biegel* 3282 (K; LISC; LMA). **Malawi**. S: Mulanje Distr., Mulanje Mt. (M'lanje) below Chambe Peak, 750 m, fl. & fr. 18.ii.1957, *J.D. Chapman* 437 (BM; K; LISC; SRGH). **Mozambique**. N: Nampula Prov., Malema, Serra Inago, 900 m, fl. & fr. 20.iii.1964, *Torre & Paiva* 11297 (COI; EA; LISC; LMA; M). Z: Ile Distr., Errego, c. 3 km from Ile Mt., 900 m, fl. 3.iii.1966, *Torre & Correia* 14968 (COI; LISC; LMA; P; PRE). MS: Manica Prov., Amatongas Forest, 600 m, fl. 1.ii.1952, *Chase* 4442 (BM). M: Marracuene, Ponta da Macaneta, c. 5 m, fl. 17.iv.1979, *Schäfer* 6737 (BM; K; LMU).

Also in Uganda, Kenya, Tanzania, Rwanda, Zaire, South Africa (KwaZulu-Natal and Transvaal), Madagascar and the Mascarene Islands, and the Yemen, India, Burma, Indo-china, Taiwan, Malesia, Micronesia, Polynesia, New Caledonia and Australia, also in N America (Florida) and S America.

In riverine and gully forests, evergreen rain forest and in woodland in mist zones, usually on moss-covered rocks in moist places and in the spray from waterfalls, also in leaf litter on forest floors and sometimes epiphytic on tree trunks; up to 2200 m.

Some material from Victoria Falls (Zambia) approaches *P. blanda* var. *blanda* in having leaves which are more than 5 cm long and which are sparsely pubescent or glabrescent. However, the mature fruits of this material lack the pseudopedicels characteristic of the typical variety.

The typical variety is glabrescent and lacks pseudopedicels. It occurs in Central and South America.

3. **Peperomia pellucida** (L.) Kunth in Humboldt Bonpland & Kunth, Nov. Gen. Sp. **1**: 64 (1816). —C. De Candolle in De Candolle, Prodr. **16**, 1: 402 (1869). —Hiern, Cat. Afr. Pl. Welw. **1**: 912 (1900). —Baker & C.H. Wright in F.T.A. **6**, 1: 149 (1909). —Hutchinson & Dalziel in F.W.T.A. **1**: 79 (1927). —Balle in Bull. Jard. Bot. État **16**: 384 (1942). —Keay, F.W.T.A., ed. 2, **1**: 82 (1954). —Düll in Bot. Jahrb. Syst. **93**: 68 et 73, fig. 1 (1973). —R.A. Howard in J. Arn. Arb. **54**: 392 (1973). —Verdcourt in F.T.E.A., Piperaceae: 11 (1996). Type from ?Martinique (see R.A. Howard, 1973).

Piper pellucidum L., Sp. Pl. **1**: 30 (1753).

Peperomia triadophylla Peter in Feddes Repert. Spec. Nov. Regni Veg. Beih. **40** [Peter, Fl. Deutsch-Ostafrika], part 2: 52, 54, & Descriptiones in Anhang: 1 (1932). Type from Tanzania.

A slender glabrous annual herb, 5–30 cm tall. Stem erect from the base, densely branched, glabrous, rarely weakly pubescent; branches divaricate. Leaves alternate, completely glabrous; petiole 4–16(18) mm long, filiform; lamina 0.5–2.7 × 0.7–3 cm, almost circular to cordate, obovate-cordate to almost reniform, rounded or sometimes broadly obtuse to acute or shortly acuminate at the apex, broadly cordate or rounded at the base, thinly membranous and ± transparent, light green on upper surface, ± glaucous or pale green beneath, palmately 5–7-nerved. Inflorescences spicate; spikes terminal and leaf-opposed, green; rhachis (0.5)2–3.6(4.2) cm long, conspicuously angled, slightly alveolate, glabrous; peduncle short, up to 0.4 cm long, filiform. Flowers rather lax on the spikes; ovary ellipsoid, crowned by a terminal penicillate stigma. Fruit 0.4–0.6 mm long, ovoid-globose, brownish, covered with longitudinal rows of minute tubercles.

Zambia. W: Kitwe, fl. & fr. 18.iii.1955, *Fanshawe* 2154 (K). C: Mkushi Distr., Lunsemfwa Wonder Gorge, W slope below van Rensburg's Saddle ("Valley View"), 970 m, fr. 18.iii.1973, *Kornaś* 3520 (K). S: South Luangwa National Park, Mfuwe Camp, fl. & fr. 23.ii.1966, *Astle* 4597 (K; SRGH). S: Mapanza West, 1070 m, fr. 7.ii.1954, *E.A. Robinson* 516 (K). **Zimbabwe**. N: E Hurungwe Distr., north bank of Maura (Mauora) R., 610 m, fr. 1.iii.1958, *Phipps* 1016 (K; LMA; SRGH). W: Hwange Distr., Lutope–Gwayi R. junction, 550 m, fr. 26.ii.1963, *Wild* 6017 (K; LISC; SRGH). E: Chipinge Distr., 200 m NNW of Musirizwi/Bwazi R. confluence, 500 m, fr. 29.i.1975, *Biegel, Pope & Russell* 4861 (K; SRGH). S: Chibi Distr., Nyoni Range, Nyuni Hill, Gororo, fr. 28.ii.1961, *Wild* 5463 (K; SRGH). **Malawi**. S: Mangochi Distr., hill 4 km NE of Mangochi, opposite Malindi turn-off, 515 m, fl. & immat. fr. 24.ii.1979, *Brummitt & Patel* 15459 (K; SRGH). **Mozambique**. N: Amaramba, Serra Mituque (Mitucué), 20 km from Cuamba (Nova Freixo), 800 m, fl. & fr. 15.ii.1964, *Torre & Paiva* 10595 (COI; LISC; SRGH). Z: Morrumbala, fr.

15.v.1943, *Torre* 5319 (BR; LISC; LMU; PRE). T: Chicoa, fl. & fr. 2.iii.1972, *Macêdo* 4973 (LISC; LMA; LMU). MS: Sofala Prov., near Sena, fr. iv.1860, *Kirk* s.n. (K).

Pantropical. Widespread in Africa from Senegal to Angola and Zaire, and from Ethiopia and the Sudan to Uganda, Kenya and Tanzania, also in Bioko and São Tomé and Principe Islands.

Moist shady places usually at river sides at low to medium altitudes, on wooded rocky hillsides and in seasonally flooded water courses and pans in woodlands, often in masses in sheltered places amongst boulders; up to 1080 m.

4. **Peperomia rotundifolia** (L.) Kunth in Humboldt Bonpland & Kunth, Nov. Gen. Sp. **1**: 65 (1816). —Brenan in Mem. New York Bot. Gard. **9**: 59 (1954). —Keay, F.W.T.A., ed. 2, **1**: 83 (1954). —Düll in Bot. Jahrb. Syst. **93**: 85 et 74, fig. 5 (1973). —Verdcourt in F.T.E.A., Piperaceae: 12 (1996). Type from ?Martinique.

Piper rotundifolium L., Sp. Pl. **1**: 30 (1753).

Peperomia bagroana C. DC. in J. Bot. **4**: 134 (1866), as "*bangroana*"; in Prodr. **16**, 1: 404 (1869), as "*bangrooana*". —Dawe, Rep. Bot. Miss. Uganda Prot.: 55 (1906), as "*bangroana*". —Baker & C.H. Wright in F.T.A. **6**, 1: 154 (1909), as "*bangroana*". —Hutchinson & Dalziel in F.W.T.A. **1**: 80 (1927). —Balle in Bull. Jard. Bot. État **16**: 388 (1942), as "*bangroana*". Type from Sierra Leone.

Peperomia mascharena C. DC. in J. Bot. **4**: 135 (1866), as "*mascarena*"; in Prodr. **16**, 1: 404 (1869), as "*mascarena*". —Baker & C.H. Wright in F.T.A. **6**, 1: 154 (1909). —S. Moore in J. Linn. Soc., Bot. **40**: 183 (1911). —Engler in Mildbraed, Wiss. Ergebn. Deutsch. Zentr.-Afr.-Exp. **2**: 177 (1911); Pflanzenw. Afrikas (Veg. Erde 9) **3**, 1: 5 (1915), as "*mascarena*". —Eyles in Trans. Roy. Soc. South Africa **5**: 337 (1916). Syntypes from South Africa (Cape Province) and Madagascar.

A creeping perennial usually epiphytic herb, up to 40 cm long. Stems prostrate, filiform, rooting at the nodes (roots very slender), sparsely pubescent. Leaves alternate; petiole 1–5 mm long, very slender; lamina 2–9(12) mm long, circular to elliptic, rounded and sometimes emarginate at the apex, cuneate at the base, rather thickly fleshy and rigid, ± sparsely puberulous on both surfaces, ciliate on the margin, palmately 3-nerved from the base, venation inconspicuous or invisible. Inflorescences of terminal or lateral greenish spikes; rhachis 5–10 mm long, erect, always shorter than the glabrous peduncle, 7–16(20) mm long. Fruit 0.5–0.7 mm long, subspheric or globose, sessile, with viscous whitish warts, and stigma subapical.

Zimbabwe. E: Chipinge Distr., Chirinda, 1060 m, fl. ii.1962, *Goldsmith* 59 (K; LISC; SRGH). **Malawi**. N: Chitipa Distr., Misuku Hills, Mughesse Forest, 1770 m, st. 20.iv.1976, *Pawek* 11150 (K; MA; MO; SRGH; UC). S: Zomba Distr., Zomba Plateau, 1400 m, fl. & fr. 28.v.1946, *Brass* 16045 (K). **Mozambique**. Z: Gurué, Serra Gurué, 3 km from waterfall of the Licungo R., 1200 m, fl. & fr. 24.ii.1966, *Torre & Correia* 14818 (COI; LISC; LMU; M; WAG). MS: Manica Prov., Serra Mocuta, c. 1100 m, fl. & fr. 3.vi.1971, *Biegel & Pope* 3524 (K; LISC; LMA; SRGH).

Pantropical. Widespread in tropical Africa, from Sierra Leone to Zaire and from Ethiopia, Uganda, Kenya and Tanzania to South Africa (Cape Province), also in Madagascar and the Comoro Islands. In medium to high altitude evergreen rain forest and riverine forests, epiphytic and often mat-forming on tree trunks, also on rocks with mosses and ferns, and in sheltered moist shady places amongst boulders; 1000–2000 m.

Düll, loc. cit. (1973) describes, and illustrates, the fruits as pseudo-pedicellate. However, pseudo-pedicels were not observed by me.

5. **Peperomia retusa** (L.f.) A. Dietr. in Sp. Pl., ed. 6, **1**: 155 (1831). —Miquel, Syst. Piper.: 132 (1843). —C. De Candolle in De Candolle, Prodr. **16**, 1: 446 (1869). —C.H. Wright in F.C. **5**, 1: 491 (1912). —Brenan in Mem. New York Bot. Gard. **9**: 59 (1954). —Keay, F.W.T.A., ed. 2, **1**: 82 (1954). —Düll in Bot. Jahrb. Syst. **93**: 89 et 75, fig. 6 (1973). —Verdcourt in F.T.E.A., Piperaceae: 13 (1996). Type from South Africa.

Piper retusum L.f., Suppl. Pl.: 91 (1782).

A perennial slightly succulent herb (4)7–30 cm tall. Stems creeping, rooting at the lower nodes, with or without leaves; branches simple or ± dichotomously branched, creeping or ascending to erect, glabrous. Leaves alternate, sometimes the uppermost opposite; petiole (1)3–8(10) mm long, glabrous, lamina (7)1–25(30) × (4)6–15(17) mm, obovate, elliptic or ovate-lanceolate, often varying in shape on the same branch, obtuse to rounded and sometimes emarginate at the apex, cuneate to broadly cuneate or almost rounded at the base, glabrous, sometimes ± pellucid punctate on both surfaces, glabrous, rarely with some scattered hairs, fleshy with a conspicuous midrib, 2–3 pairs of lateral nerves and a marginal nerve, venation scarcely visible or lacking; margins ciliate particularly toward the apex. Inflorescences spicate; spikes terminal or axillary, solitary or rarely 2–3-clustered, greenish; rhachis (5)10–40(50) mm long, glabrous, slightly thicker than the glabrous (3)4–15 mm long peduncle. Ovary ovoid ± sunk into the rhachis; bracts glandular-punctate. Fruit 0.4–0.7 mm long, ovoid to spheric, bright brown, covered with whitish viscous glands, shortly pseudopedicellate, with a subapical stigma.

1. Plants small; leaves usually not exceeding 10(15) mm in length; rhachis 5–15 mm long; peduncle 4–10 mm long ·· var. *bachmannnii*
- Plants larger; leaves mostly 10–30 mm long; rhachis 8–50 mm long; peduncle 5–15 mm long ·· 2
2. Leaves usually obovate up to 15 mm long, obtuse or rounded and emarginate at the apex, cuneate at the base; creeping part of stem almost devoid of leaves; peduncle 5–15 mm long; rhachis 8–30 mm long ································ var. *retusa*
- Leaves usually ovate, elliptic or ovate-lanceolate, reaching 30 mm long, obtuse at the apex, broadly cuneate at the base; stems leafy for most of their length; peduncle usually more than 10 mm long; rhachis 20–50 mm long ·························· var. *mannii*

Var. **retusa**
  Peperomia retusa var. *alternifolia* C. DC. in De Candolle, Prodr. **16**, 1: 446 (1869). Type from South Africa.
  Peperomia retusa var. *ciliolata* C. DC. in De Candolle, Prodr. **16**, 1: 447 (1869). —C.H. Wright in F.C. **5**, 1: 491 (1912). Type from South Africa.
  Peperomia rehmannii C. DC. in Bot. Jahrb. Syst. **19**: 227 (1894). Type from South Africa.

Zambia. E: Chama Distr., Makutus, fl. 29.x.1972, *Fanshawe* 11632 (K). **Zimbabwe**. E: Bikita Distr., Mt. Horzi, 1275 m, fl. 11.v.1969, *Biegel* 3123 (K; SRGH). **Malawi**. N: Chitipa Distr., Misuku Hills, Mughesse Forest, 1800 m, fl. & fr. 1.iii.1983, *Dowsett-Lemaire* 638 (K). S: Mt. Mulanje (M'lanje), Big Ruo Gorge, fl. & fr. 7.vii.1958, *J.D. Chapman* H676 (SRGH). **Mozambique**. MS: Sofala Prov., Serra Gorongosa, Mt. Nhandore, 1800 m, fl. 21.x.1965, *Torre & Pereira* 12499 (LISC; LMA; PRE).
  Also in Zaire, Uganda, Kenya, Tanzania and South Africa (KwaZulu-Natal, Transvaal and Cape Province). Submontane evergreen rainforest and gully forest, epiphytic on tree trunks and amongst mosses on rocks boulders and fallen trees, forming small colonies in moist shady places; 330–1900 m.

Var. **mannii** (Hook.f.) Düll in Bot. Jahrb. Syst. **93**: 90 et 75, fig. 6, a1–a3 (1973). Type from Cameroon.
  Peperomia mannii Hook.f. in J. Linn. Soc., Bot. **7**: 217 (1864). —C. De Candolle in De Candolle, Prodr. **16**, 1: 422 (1869). —Baker & C.H. Wright in F.T.A. **6**, 1: 153 (1909). —Hutchinson & Dalziel in F.W.T.A. **1**: 80 (1927).
  Peperomia mannii var. *fernandopoana* C. DC. in De Candolle, Prodr. **16**, 1: 422 (1869). —Baker & C.H. Wright in F.T.A. **6**, 1: 153 (1909). Type from Bioko.
  Peperomia bueana C. DC. in Bot. Jahrb. Syst. **19**: 227 (1894). —Baker & C.H. Wright in F.T.A. **6**, 1: 153 (1909). —Hutchinson & Dalziel, loc. cit. (1927). Type from Cameroon.
  Peperomia usambarensis Engl., Bot. Jahrb. Syst. **45**: 276 (1910); Pflanzenw. Afrikas (Veg. Erde 9) **3**, 1: 5 (1915). Type from Tanzania.
  Peperomia gracilipetiolata De Wild. in Rev. Zool. Bot. Africaines **8**, Suppl. Bot.: 8 (1920). Type from Uganda/Zaire.

Zimbabwe. E: Mutare Distr., Engwa, 1830 m, fl. & fr. 3.ii.1955, *Exell, Mendonça & Wild* 168 (BM; LISC; SRGH). S: Mberengwa Distr., Mt. Buhwa, c. 1200 m, fl. & fr. 2.v.1973, *Biegel, Pope & Simon* 4265 (K; SRGH). **Malawi**. N: Nyika Plateau, 2250 m, fl. & fr. 16.viii.1946, *Brass* 17239 (K; SRGH). S: Mt. Mulanje (M'lanje), Little Ruo Plateau, 1750 m, fl. & fr. 17.viii.1956, *Newman & Whitmore* 482 (BM; SRGH). **Mozambique**. Z: Milange, Serra Chiperone, hillside NW, 1600 m, fl. & fr. 1.ii.1972, *Correia & Marques* 2476 (LMU). MS: Manica Prov., Vumba, fl. 23.vi.1949, *Pedro & Pedrogoã* 6865 (LMA).
  Also in West Tropical Africa, Cameroon, Bioko and São Tomé, and in Zaire, Uganda, Kenya and Tanzania. Mixed evergreen forest, rain forest and gully forest, epiphytic on trees and fallen tree trunks, shaded damp moss-covered boulders and on soil and rocks beside waterfalls; 900–2500 m.

Var. **bachmannii** (C. DC.) Düll in Bot. Jahrb. Syst. **93**: 90 et 75, fig. 6, b1–b5 (1973). Type from South Africa.
  Peperomia bachmannii C. DC. in Bot. Jahrb. Syst. **19**: 227 (1894). —C.H. Wright in F.C. **5**, 1: 491 (1912). —Engler, Pflanzenw. Afrikas (Veg. Erde 9) **3**, 1: 5 (1915).

Peperomia wilmsii C. DC. in Annuaire Conserv. Jard. Bot. Genève **2**: 282 (1898). Type from South Africa.
Peperomia ulugurensis Engl., Bot. Jahrb. Syst. **28**: 374 (1900). —Baker & C.H. Wright in F.T.A. **6**, 1: 151 (1909). —Weimarck in Bot. Notis. **1936**: 1 (1936). —Balle in Bull. Jard. Bot. État **16**, 1: 390 (1942). Type from Tanzania.
Peperomia ukingensis Engl., Bot. Jahrb. Syst. **30**: 289 (1901); in Mildbraed, Wiss. Ergebn. Deutsch. Zentr.-Afr.-Exp. **2**: 177 (1911); Pflanzenw. Afrikas **3**, 1: 5, fig. 3B (1915). Type from Tanzania.
Peperomia subdichotoma De Wild. in Rev. Zool. Bot. Africaines **8**, Suppl. Bot.: 9 (1920). Type from Zaire.
Peperomia ulugurensis var. *acutifolia* Balle in Bull. Jard. Bot. État **16**, 1: 391 (1942). Type from Zaire.
Peperomia ulugurensis var. *diversifolia* Balle in Bull. Jard. Bot. État **16**, 1: 391 (1942). Type from Zaire.
Peperomia ulugurensis var. *ukingensis* (Engl.) Balle in Bull. Jard. Bot. État **16**, 1: 391 (1942). Type as for *Peperomia ukingensis*.

Zimbabwe. E: Mutare Distr., Vumba, Norseland, 1500 m, fl. & fr. iii.1949, *Wild* 2836 (K; SRGH). **Malawi**. N: Chitipa Distr., Misuku Hills, Mughesse Forest, 1680 m, fl. 1.i.1977, *Pawek* 12162 (K; MO). S: Mt. Mulanje (Mlanje), Tuchila Plateau, 1830 m, *Newman & Whitmore* 211 (BM).
Also in Zaire, Rwanda, Uganda, Kenya, Tanzania, and South Africa (Transvaal, KwaZulu-Natal and Cape Province). Submontane evergreen rain forest and gully forest, epiphytic on trees and on shaded damp moss-covered rocks; 985–2750 m.

6. **Peperomia goetzeana** Engl., Bot. Jahrb. Syst. **28**: 375 (1900). —Baker & C.H. Wright in F.T.A. **6**, 1: 152 (1909). —Brenan in Mem. New York Bot. Gard. **9**, 1: 59 (1954). —Düll in Bot. Jahrb. Syst. **93**: 99 et 77, fig. 10 (1973). Type from Tanzania.
Peperomia rungwensis Engl., Bot. Jahrb. Syst. **30**: 290 (1901). Type from Tanzania.

A succulent perennial herb 15–50 cm tall, little-branched, glabrous. Stems decumbent, with internodes c. 20 mm long and 3–4 mm in diameter, rooting at the nodes, rather leafy at the erect distal end. Leaves alternate, sometimes the uppermost opposite; petiole (2)5–8(10) mm long, thick; lamina (22)30–50(65) × (15)19–28(37) mm, spathulate to obovate-oblong, varying markedly on the same branch, rounded to narrowed and obtuse or sometimes emarginate at the apex, broadly cuneate at the base, succulent when fresh, bright green on upper surface, pale green beneath, drying coriaceous to membranaceous and ± dark coloured, pellucid glandular-punctate, 3–5-nerved, with the midrib very prominent below, glabrous, ± ciliate on the margins towards the apex. Inflorescences spicate; spikes solitary, terminal or axillary, greenish; rhachis (3)4–6.5(8.5) cm long, 0.2–0.3 cm in diameter; peduncle (0.5)1.5–2.5(3.0) cm long. Ovary subglobose, stigma subapical; bracts 0.5–0.7 mm in diameter, subcircular, peltate, glandular-punctate. Fruit c. 1 mm in diameter when mature, spheric-ellipsoid, brownish, covered with whitish glands; stigma subapical.

Malawi. N: Rumphi Distr., Nyika Plateau, Kasaramba Peak, 2560 m, fl. & fr. 9.ii.1968, *Simon, Williamson & Ball* 1727 (K; LISC; SRGH). C: Dedza Distr., Dedza Mt., fl. i.1961, *J.D. Chapman* 1103 (K; SRGH). **Mozambique**. MS: Manica Prov., Serra Mocuta, 750 m, fl. & fr. 3.vi.1971, *Pope* 442 (K; LISC; SRGH).
Also in the Sudan, Ethiopia, Kenya, Uganda and Tanzania. In medium to high altitude montane evergreen rainforests, on rocks and boulders with mosses, often beside streams, also as an epiphyte on forest tree trunks; 750–3000 m.

7. **Peperomia molleri** C. DC. in Bol. Soc. Brot., sér. 1, **10**: 154 (1892). —Baker & C.H. Wright in F.T.A. **6**, 1: 150 (1909). —Exell, Cat. Vasc. Pl. S. Tomé: 277 (1944). —Keay, F.W.T.A., ed. 2, **1**: 82 (1954). —Düll in Bot. Jahrb. Syst. **93**: 106 et 79, fig. 15 (1973). —Verdcourt in F.T.E.A., Piperaceae: 16 (1996). Type from São Tomé.
Peperomia fernandopoiana var. α C. DC. in J. Bot., London **4**: 134 (1866). Syntypes from Angola and São Tomé.
Peperomia fernandopoiana var. *subopacifolia* C. DC. in De Candolle, Prodr. **16**: 397 (1869), as *"fernandopoana"*. Syntypes as above.
Peperomia holstii C. DC. in Bot. Jahrb. Syst. **19**: 226 (1894). —Engler, Pflanzenw. Ost-Afrikas **C**: 159 (1895). —Hiern, Cat. Afr. Pl. Welw. **1**: 913 (1900). —Baker & C.H. Wright in F.T.A. **6**, 1: 151 (1909). —Engler, Pflanzenw. Afrikas (Veg. Erde 9) **3**, 1: 5 (1915). Type from Tanzania.

Peperomia holstii var. *elongata* De Wild. in Ann. Mus. Congo Belge, Bot., sér. 5, **3**: 62 (1909). —Balle in Bull. Jard. Bot. État **16**: 398 (1942). Type from Zaire.
Peperomia magilensis Baker in F.T.A. **6**, 1: 150 (1909). Type from Tanzania.
Peperomia stolzii C. DC. in Bot. Jahrb. Syst. **57**: 19 (1920). Type from Tanzania.

A somewhat succulent glabrous perennial herb up to 30(50) cm tall. Stems stoloniferous in the lower part and rooting at the lower nodes; branches slender, narrowly winged, ascending. Leaves alternate, sometimes the uppermost opposite; petiole (4)6–15(20) mm long; lamina (2)2.5–6(9) × (1)1.5–3.5(4.5) cm, broadly ovate to broadly lanceolate, acute to ± acuminate or sometimes obtuse at the apex, never emarginate, broadly cuneate to rounded at the base, dark green and papery when dry, sometimes obscurely bright pellucid-punctate, glabrous on both surfaces, ciliate towards the apex, 3–5(7)-nerved from the base; midrib prominent below, lateral venation scarcely visible or absent. Inflorescence spikes solitary, terminal or rarely the uppermost axillary, greenish; rhachis (2.5)3.2–7.0(10) cm long, slender and pitted; peduncle (1.1)2–3(6) cm long, slender. Ovary ovoid, sessile; stigma subapical; bracts 0.3–0.5 mm in diameter, circular, glandular-punctate. Fruit c. 0.1 cm spheric-ellipsoid, erect, brown, covered with viscous glands, stigma subapical.

Zimbabwe. E: Chipinge Distr., Chiredza Gorge, c. 915 m, fl. & fr. iii.1962, *Goldsmith* 65 (SRGH).
Also in West Tropical Africa, Cameroon, Bioko and São Tomé, and in Angola, Zaire, Burundi, Uganda, Kenya and Tanzania. Evergreen rain forest and gully forest, usually on forest floor but sometimes on moss-covered rocks or epiphytic on trees; 110–2090 m.

8. **Peperomia abyssinica** Miq. in Hooker, London J. Bot. **4**: 419 (1845). —Baker & C.H. Wright in F.T.A. **6**, 1: 153 (1909). —Fries, Wiss. Ergebn. Schwed. Rhod.-Kongo-Exped. **1**, 1: 11 (1914). —Engler, Pflanzenw. Afrikas (Veg. Erde 9) **3**, 1: 5 (1915). —Balle in Bull. Jard. Bot. État **16**: 396 (1942). —Brenan in Mem. New York Bot. Gard. **9**, 1: 59 (1954). —Verdcourt in F.T.E.A., Piperaceae: 15 (1996). Type from Ethiopia.
Peperomia stuhlmannii C. DC. in Bot. Jahrb. Syst. **19**: 225 (1894). —Engler, Pflanzenw. Ost-Afrikas **C**: 159 (1895). —Baker & C.H. Wright in F.T.A. **6**, 1: 151 (1909). Type from Uganda.
Peperomia emarginata De Wild. in Rev. Zool. Bot. Africaines **8**, Suppl. Bot.: 9 (1920). Type from Zaire.
Peperomia goetzeana sensu Verdcourt in F.T.E.A., Piperaceae: 15 (1996).

A succulent completely glabrous perennial herb. Stems creeping and stoloniferous in the lower part, rooting at the lower nodes, and without leaves, becoming erect to 25 cm tall and leafy in the upper part, with thickened nodes. Leaves alternate; petiole (2)3–5(10) mm long, thickened; lamina (1.8)2.2–4.7(5.5) × (0.7)1.0–1.5(2) cm, rhomboid-elliptic, elliptic, sometimes ovate, rarely obovate in lower leaves, somewhat acuminate or subacute to obtuse at the apex, sometimes emarginate, broadly cuneate at the base, ± fleshy, chartaceous when dry, glandular-punctate on both surfaces, ciliate near the apex, 3–5-nerved from the base with midrib conspicuous. Inflorescence spikes usually solitary, terminal or axillary; rhachis 2–4(9) cm long, c. 0.2 mm in diameter, rarely 2-forked; peduncle (0.5)0.7–1.5(2) cm long. Ovary ovoid, stigma subapical; bracts (0.5)0.7 mm in diameter, circular, glandular-punctate. Fruit c. 0.9 mm long, spheric-ovoid, covered with glandular warts, shortly pseudopedicellate, stigma subapical.

Var. **abyssinica** —Düll in Bot. Jahrb. Syst. **93**: 95 & 76, fig. 8a (1973).

Malawi. N: Rumphi Distr., Nyika Plateau, Chosi Forest, 2300 m, fl. 25.xii.1975, *Phillips* 729 (K). C: Nkhota Kota Distr., Ntchisi Mt., 1550 m, fl. 26.vii.1946, *Brass* 16962 (BM; K; SRGH). S: Mt. Mulanje (Mlanje) from Sombani to Great Ruo Basin, fl. 1.i.1971, *Hilliard & Burtt* 6088 (LMU).
Also in Ethiopia, Somalia, Kenya, Uganda, Tanzania and Zaire. Submontane evergreen rain forests, relic *Widdringtonia* patches and *Brachystegia* woodland bordering evergreen forests, on forest floor and on moss-covered rocks in forests and stream beds, or in sheltered places amongst rocks in woodland, also sometimes epiphytic on tree trunks; 1550–2850 m.
P. abyssinica var. *stuhlmannii* (C. DC.) Düll, occurs in Kenya, Uganda and Tanzania, and may be distinguished from the typical variety in the following features: petiole. c. 0.5 cm long; leaves c. 5.5 cm long, elliptic-lanceolate not emarginate; peduncle 2–3 cm long; rhachis c. 5 cm long and fruit c. 1.2 mm long.

Species insufficiently known

Dowsett-Lemaire 669 (K) and *Pawek* 13064 (K), both from the Mughesse Forest in the Misuku Hills of Malawi, and both with pubescent stems and leaves, approach *P. blanda* var. *leptostachya* in indumentum and venation but differ from it in having alternate leaves. The leaf shape of this material is similar to that of *P. goetzeana* but that species is glabrous. Fertile material is needed before the affinities of this taxon can be established.

143. MYRISTICACEAE

By G.V. Pope

Trees or shrubs, rarely lianes, evergreen or rarely deciduous, often aromatic, dioecious or monoecious; sap coloured. Leaves alternate, simple, entire, often gland-dotted; exstipulate. Indumentum of stellate-branching, shortly-stalked T-shaped or uniseriate hairs. Inflorescences paniculate, fasciculate-racemose, or capitate, sometimes apparently cymose; bracts often present, usually deciduous, bracteoles mostly absent. Flowers unisexual, small, actinomorphic, whitish-yellow, pink or red; perianth saucer- or funnel-shaped, campanulate or urceolate, (2)3–4(5)-valvately lobed. Male flowers with 2–30(40) stamens; filaments partially or completely fused into a column; anthers laterally connate, rarely free, 2-thecous, dehiscing by longitudinal slits; rudimentary ovary absent. Female flowers 1-carpellate; ovary superior, usually sessile; style small or absent; stigmas ± 2-lobed; ovule 1, usually anatropous. Fruits a drupe with a fleshy or leathery pericarp, mostly dehiscing into 2 valves. Seeds partially or completely enveloped by an often brightly coloured laciniate or subentire fleshy aril; testa usually in 3 layers, the outer membranous or fleshy, the middle one woody and the inner one membranous, usually intruding into the folds of the endosperm; endosperm abundant, entire or ruminate, very oily; embryo very small with basal suberect, spreading or connate cotyledons.

A pantropical family of 16–19 genera and 240–400 species. Of the 5 genera recorded for Africa two occur naturally in the Flora Zambesiaca area.

Myristica fragrans Houtt is the source of the spices "nutmeg" (the seed) and "mace" (the dried arils). Originally native of the Moluccas it is now grown throughout the tropics. Introduced plants have been cultivated on the islands of Zanzibar and Pemba and in the East Usambara mountains of Tanzania, but have so far not been recorded from the Flora Zambesiaca area.

Leaves ± cordate at the base; main lateral nerves pronounced, in 15–60 pairs; tertiary nerves scalariform; hairs of the indumentum branching-stellate, ± erect; aril deeply laciniate; endosperm ruminate · 1. **Pycnanthus**
Leaves rounded to broadly cuneate at the base; main lateral nerves faint, in 5–10 pairs; tertiary nerves reticulate; hairs of the indumentum shortly-stalked T-shaped, ± appressed; aril almost completely enveloping the seed, lobulate at apical rim; endosperm not ruminate · 2. **Staudtia**

1. PYCNANTHUS Warb.

Pycnanthus Warb. in Notizbl. Königl. Bot. Gart. Berlin **1**: 99 (1895); in Nova Acta Acad. Caes. Leop.-Carol. German. Nat. Cur. **68**: 130, 252 (1897). —Kühn & Kubitzki, Fam. Gen. Vasc. Pl. **2**: 464, fig. 99C (1993).

Evergreen trees or lianes, dioecious or monoecious. Leaves mostly cordate at the base, minutely punctate and ± glaucous beneath, strongly pinnately nerved; lateral nerves prominent, numerous, tertiary venation subparallel. Inflorescences paniculate, divaricately branched with flowers clustered in numerous dense obovoid-globose heads at the ends of the inflorescence branches; flowers sessile; bracteoles absent. Male flower perianth 3–4-lobed; stamen filaments fused into a slender column; anthers 2–6, adnate to the column about the apical part. Female flower perianth as in male; stigma sessile, indistinct. Fruit oblong, ellipsoid or ± globular with a thick cartilaginous pericarp, dehiscent; pericarp thickly woody-crustaceous in mature fruits. Seed testa somewhat thin; aril laciniate; endosperm fatty, ruminate; cotyledons free almost to the base, suberect.

A genus of 3–4 species in tropical Africa.

Pycnanthus angolensis (Welw.) Warb. in Notizbl. Königl. Bot. Gart. Berlin **1**: 100 (1895). — Exell, Cat. Vasc. Pl. São Tomé: 278 (1944). —Gilbert & Troupin in F.C.B. **2**: 391 (1951). — Gossweiler in Agronomia Angolana No.7: 397 (1953). —Keay in F.W.T.A. ed. 2, **1**: 61 (1954). —Fouilloy in Fl. Gabon, No. 10: 87, pl. 22 (1965); in Fl. Cameroun **18**: 91, pl. 27 (1974). TAB. **11**. Type from Angola.
Myristica sp. —Welwitch in Annaes Conselho Ultram. [Apontamentos Phytogeographicos] **1**: 554 (1859).
Myristica angolensis Welw., Synop. Explic. Mad. et Drog. Med.: 51 (1862). —Hiern, Cat. Afr. Pl. Welw. **1**, part 4: 913 (1900).
Myristica kombo Baill., Adansonia **9**: 79 (1868). Type from Gabon.
Pycnanthus microcephalus sensu Warb. in Ber. Pharm. Ges. Berl. **1892**: 222, 226 (1892) pro parte quoad *Myristica angolensis* Welw. sed non *Myristica microcephala* Benth. (1878).
Pycnanthus kombo (Baill.) Warb. in Nova Acta Acad. Caes. Leop.-Carol. German. Nat. Cur. **68**: 252, tabs 2 & 10 (1897) pro parte quoad specim. *Welwitsch* 581; Muskatnuss: 374, 385, t. 4, fig. 2 (1897). —Stapf in F.T.A. **6**, 1: 158 (1909). —Hutchinson in F.W.T.A. ed. 1, **1**: 64 excl. fig. 15 (1927). —De Wild., Contrib. Fl. Katanga, Suppl. 2: 4 (1929).
Pycnanthus mechowii Warb. in Nova Acta Acad. Caes. Leop.-Carol. German. Nat. Cur. **68**: 261, t. 10, figs. 1, 2 (1897). —Stapf in F.T.A. **6**, 1: 160 (1909). Type from ?Zaire.
Pycnanthus kombo var. *angolensis* (Welw.) Warb. in Nova Acta Acad. Caes. Leop.-Carol. German. Nat. Cur. **68**: 257 (1897), excl. specim. *Stuhlmann* 1167.

Evergreen tree up to 22(37) m tall; bole up to 0.7(1.5) m d.b.h., branching high up to produce a small crown; sap copious viscous yellow, turning dark red on exposure to air; branches patent, branchlets slender, ± pendulous at the tips, young growth densely tomentose, the hairs reddish-brown stellate and easily rubbed off. Leaves 7.5–31 × 4.3–11 cm, oblong to oblong-lanceolate, or ± elliptic, sometimes oblong-oblanceolate, ± long acuminate to subcaudate at the apex, cordate or subcordate at the base, dark green on upper surface, glaucous and very densely minutely glandular-punctate beneath, glabrous or with scattered dark brown stellate hairs on the lower surface more numerous on midrib especially when young; lateral nerves 15–20(40), prominent below; petiole 7–14 mm long, dark brown tomentose with caducous stellate hairs. Inflorescences usually borne on 2- or 3-year old branches, below the leaves, 10–15 cm long, paniculate with flowers many in numerous dense clusters on short lateral branches, densely rusty tomentose; bracts subtending the individual clusters of heads of flowers 2–2.5 × 5 mm, tomentose, deciduous. Flowers unisexual, scented, the perianth completely covered with dark brown vesiculose hairs; plants dioecious. Male flowers: perianth c. 1 mm long, deeply 3(5)-lobed; lobes obovate; staminal column 1–1.2 mm long, sometimes twisted; anthers 2–4, exserted on the elongating staminal column, the panicle falling entire after flowering. Female flowers: perianth lobes broadly ovate; ovary ± 0.5 mm in diameter. Infructescence pendulous, branches becoming stout; fruits subsessile, clustered, 3–4(4.5) × (1.6)2–3(3.8) cm, ellipsoid to oblong or pyriform, splitting vertically into two halves, dark brown tomentose when young, yellowish-orange when ripe; pericarp cartilaginous, 2–10 mm thick, becoming thickly woody-crustaceous when dry. Seed single, 1.5–2.9 × 0.8–1.5 cm, ellipsoid-oblong, dark brown. Aril fleshy, laciniate almost to the base, red or pink.

Subsp. **schweinfurthii** (Warb.) Verdc., comb. et stat. nov.* Type from Zaire, Assika Stream, *Schweinfurth* (K, *Economic Bot. Coll.* 45471, lectotype; B†; K, paralectotypes).

---

* Verdcourt, in F.T.E.A. manuscript, comments as follows: "Infraspecific variation in *Pycnanthus angolensis* is far from clear but I have concluded that the NE African population with mostly more globose fruits and thicker pericarp is distinct at some level and maintain it as a subspecies; how far it extends to the west is uncertain. Material is inadequate from Angola to obtain a complete picture of variation there".

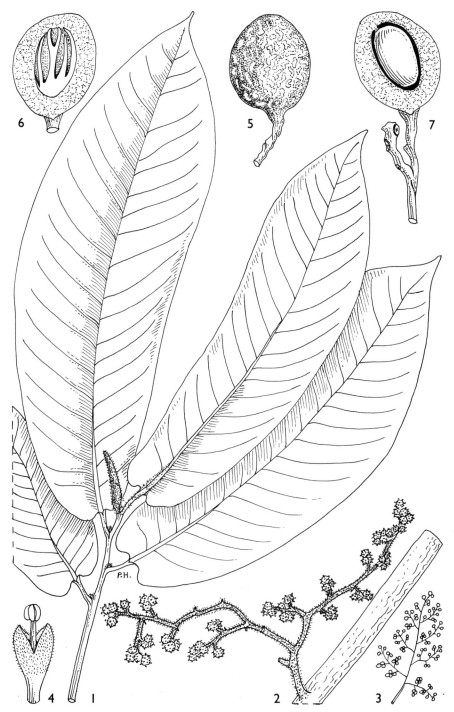

Tab. 11. PYCNANTHUS ANGOLENSIS. 1, distil portion of leafy branch (×²/₃), from *Pirozynski* 361; 2, male inflorescence, with capitate flower clusters (×²/₃); 3, inflorescence, diagrammatic; 4, male flower (×18), 2–4 from *Katende* 210; 5, fruit (×²/₃); 6, fruit with 1 valve removed, showing thick pericarp and seed with aril (×²/₃); 7, fruit showing thick pericarp, seed without aril and woody peduncle (×²/₃), 5–7 from *Snowden* 1781. Drawn by Pat Halliday.

Pycnanthus schweinfurthii Warb: in Engl., Pflanzenw. Ost-Afrikas **B**: 271 (1895); in
Pflanzenw. Ost-Afrikas **C**: 180 (1895); in Nova Acta Acad. Caes. Leop.-Carol. German. Nat.
Cur. **68**: 260, t. 10, figs. 1–3 (1897). —Stapf in F.T.A. **6**, 1: 160 (1909).
Pycnanthus kombo var. *angolensis* sensu Warb. in Nova Acta Acad. Caes. Leop.-Carol.
German. Nat. Cur. **68**: 257 (1897) pro parte quoad *Stuhlmann* 1167 non (Welw.) Warb.
Pycnanthus kombo var. *sphaerocarpus* Stapf in F.T.A. **6**, 1: 159 (1909). Type from Tanzania.
Pycnanthus sp. 1 (*White* 3637), F. White, F.F.N.R.: 56 (1962).

Fruits mostly 3.3–4 × 2–3(4) cm, oblong-ellipsoid to globose; pericarp 4–10 mm
thick; seeds (1.5)2–2.8 × 1.2–1.6 cm. Adult leaves ± glabrous.

Zambia. N: near Kawambwa Boma, 8.xi.1952, *White* 3637 (K; FHI); Mbala Distr., new road to
Chinakila from Kambole road, banks of Ulwala R., 4.x.1967, *Richards* 22342 (K).
    Also in the Sudan, Uganda, Burundi Tanzania and Zaire. Infrequent canopy tree of swamp
and riverine forest; 1200–1400m.
    Subsp. *angolensis* occurs in Angola and W Africa, and perhaps in Zaire. Its fruits are smaller
and have a thinner pericarp.

2. STAUDTIA Warb.

Staudtia Warb. in Nova Acta Acad. Caes. Leop.-Carol. German. Nat. Cur. **68**: 128
(1897). —Kühn & Kubitzki, Fam. Gen. Vasc. Pl. **2**: 464 (1993).

Evergreen trees, monoecious or dioecious. Leaves subcoriaceous, rounded to
cuneate at the base, minutely punctate beneath, lateral nerves not prominent,
tertiary venation reticulate. Inflorescences capitate with flowers densely clustered in
sub-globose shortly pedunculate heads in leaf axils, bracteate; bracteoles absent;
flowers unisexual, shortly pedicellate. Male flower perianth funnel-shaped, 3–4-
lobed; stamen filaments fused into a slender column; anthers 3–4, adnate to the
column. Female flower perianth as in male; stigma sessile. Fruits clustered on a
shortly-stalked receptacle, pedicellate, ovoid, with a thick cartilaginous pericarp,
dehiscent, the suture sometimes raised into a low ridge; pericarp thickly woody-
crustaceous in mature fruits. Seeds ellipsoid; aril ± lobed at the apex, or cupular and
entire; endosperm not ruminate.

A tropical African genus, with 2 species.

Staudtia kamerunensis Warb. in Nova Acta Acad. Caes. Leop.-Carol. German. Nat. Cur. **68**: 129,
    241 (1897). —Stapf in F.T.A **6**, 1:165 (1909). —Keay in F.W.T.A. ed. 2, **1**: 62 (1954). Type
    from Cameroons.
        Staudtia gabonensis var. *macrocarpa* Gilbert & Troupin in F.C.B. **2**: 396 (1951); in Bull.
    Jard. Bot. État **21**: 142 (1951). Type from Zaire.
        Myristica niohue Baill., Adansonia, ser. 1, **9**: 79 (1868) pro parte. Type from Gabon.

Var. **gabonensis** (Warb.) Fouilloy in Fl. Cameroun **18**: 104, pl. 31, figs. 1–7 (1974). TAB. **12**.
    Type from Gabon.
        Staudtia gabonensis Warb. in Bot. Jahrb. Syst. **33**: 384 (1903). —Stapf in F.T.A. **6**, 1: 165
    (1909). —Vermoesen, Man. Ess. Forest. Congo Belge: 255, cum pl. (1923). —Gilbert &
    Troupin in F.C.B. **2**: 394 (1951). —Gossweiler in Agronomia Angolana No. 7: 398 (1953).
    —Fouilloy in Fl. Gabon, No. 10: 99, pl. 25, figs. 1–7 (1965).
        Staudtia stipitata Warb. in Bot. Jahrb. Syst. **33**: 384 (1903). —Stapf in F.T.A. **6**, 1: 165
    (1909). —Keay in F.W.T.A. ed. 2, **1**: 62 (1954). Type from Cameroons.
        Staudtia congensis Vermoesen Man. Ess. Forest. Congo Belge: 257 (1923). Type from
    Congo.

Evergreen tree 10–28(35) m tall; bole slender, up to 0.75 m d.b.h., narrowly
buttressed to 2 m; bark rough, flaking in circular patches to leave brown scars; slash
red with abundant red, somewhat stringy latex; branches high up, patent spreading;
branchlets glabrous, smooth, yellowish-brown; buds and young growth at first
appressed pubescent, the indumentum reddish-brown consisting of short-stalked ± T-
shaped hairs, easily rubbed off. Leaves subcoriaceous, 7–15(18) × 2.2–6.5 cm,
oblong to oblong-lanceolate, or ± narrowly elliptic, sometimes oblong-oblanceolate,
long acuminate to caudate at the apex or sometimes obtuse, broadly cuneate to ±

Tab. 12. STAUDTIA KAMERUNENSIS var. GABONENSIS. 1, fertile twig (×²⁄₃), from *Eggeling* 4253; 2, male inflorescence (×2); 3, male flower (×10); 4, longitudinal section through male flower (×10), 2–4 after *Fouilloy*; 5, female inflorescence (×2); 6, female flower (×10), 5 & 6 from *Tailfer-Patambalu* 4; 7, infructescence, mature (×²⁄₃), from *Kennedy* 2337; 8, single mature fruit (×²⁄₃); 9, aril with seed (×²⁄₃); 10, aril split open (×²⁄₃), 8–10 from *Eggeling* 4253 spirit material; 11, T-shaped hairs from buds and young growth, much enlarged, after *Fouilloy*. Drawn by Pat Halliday.

rounded at the base, dark green above, paler and very densely minutely glandular-punctate beneath, glabrous; margins entire; main lateral nerves in 5–10 pairs, ± obscure above, slightly raised below, tertiary venation reticulate; petiole 5–17 mm long, dark brown pubescent with caducous appressed hairs. Inflorescences usually borne on 2-year old branches, usually in axils of fallen leaves, subglobose-capitate; peduncle 3–5 mm long, swelling into a club-shaped apex, the lower part resembling the twig and several–many-bracteate, the bracts broadly ovate and hairy. Male inflorescence 4–6 mm in diameter; flowers 20–50 on a subglobose receptacle, flower and pedicel up to c. 2 mm long, funnel-shaped, 3-lobed, ferruginous-puberulous; anthers 3, adnate to apex of staminal column. Female inflorescence 6–9 mm in diameter; flowers 10–30 on a subglobose receptacle, 3-lobed, ferruginous-puberulous, flower and pedicel c. 3 mm long, narrowly funnel-shaped; pedicels elongating in fruit; ovary ± globose, stigma sessile. Infructescence stalk (peduncle) 3–8 mm long, stout, woody; fruits on pedicels 3–8 mm long and 3–4 mm wide (occasionally also stipitate at the base), in clusters of 2–20, red and somewhat fleshy when ripe, 13–25(36) × 15–19 mm, ellipsoid to ovoid, splitting vertically into two valves, dark brown puberulous when young, glabrescent; pericarp cartilaginous, up to c. 3 mm thick, becoming thickly woody-crustaceous when dry. Seed single, 12–20 × 8–12 mm, narrowly ovoid, dark brown. Aril enveloping $^3/_4$ or more of the seed, lobulate at the apex, thinly fleshy, red or pink; lobes 2–5 mm long and c. 1 mm wide, shortly strap-shaped and usually bifid at the apex.

Zambia. W: Mwinilunga Distr., source of Matonchi R., 16.ii.1938, *Milne-Redhead* 4594 (K); Mwinilunga, fr. 12.x.1955, *W.D. Holmes* 1261 (K; NDO); Solwezi Distr., Kifubwe Gorge, fr. 15.v.1969, *Mutimushi* 3456 (K; NDO).

Also in Nigeria, Cameroons, Gabon, Cabinda, Angola, Zaire, Central African Republic, Uganda and Burundi. Evergreen riverine forest and margins of swamp forest (mushitu).

S. kamerunensis var. *kamerunensis* occurs in Cameroons and Zaire, and has larger fruits, 45 × 45 mm.

144. MONIMIACEAE

By B.L. Stannard

Trees, shrubs or rarely lianes, often aromatic. Leaves opposite or less often alternate, exstipulate, simple, entire or toothed, sometimes glandular-toothed, coriaceous, sometimes minutely pellucid-punctate. Inflorescences axillary or terminal, racemose, cymose or fasciculate; bracts small or absent. Flowers hermaphrodite or unisexual, actinomorphic or more rarely zygomorphic; plants monoecious, sometimes dioecious, less often polygamous. Perianth 4–many-lobed; lobes in 2–several whorls, imbricate, rarely absent. Stamens few to numerous, 1–2-seriate or irregular, usually free; filaments very short, sometimes with glands at the base; anthers erect, 2-locular, dehiscing by longitudinal slits, or more rarely by valves. Staminodes sometimes present in female flowers. Ovary superior; carpels usually several, rarely solitary, 1-locular; stigma sessile or with a long or short style; ovules solitary, erect or pendulous, usually anatropous. Fruit of separate drupes, or more rarely achenes, often enclosed by the perianth or in a fleshy receptacle. Seeds erect or pendulous, endosperm fleshy, copious.

A family of 34 genera and c. 440 species in the tropics and subtropics, predominantly in the southern hemisphere. Two genera, *Xymalos* and *Glossocalyx*, occur in Africa. Only *Xymalos* is found in the Flora Zambesiaca area.

XYMALOS Baill.

Xymalos Baill. in Bull. Mens. Soc. Linn. Paris 1: 650 (1887). —W.R. Philipson in Kubitzki et al., Fam. Gen. Vasc. Pl. 2: 426 (1993).

Paxiodendron Engl., Pflanzenw. Ost-Afrikas C: 182 (1895).

Small trees or shrubs, glabrous except for the inflorescence. Leaves opposite or subopposite. Inflorescences axillary, solitary or paired racemes or panicles, yellowish- or greyish-tomentose; bracts ovate to triangular. Flowers very small. Male flowers: perianth 4–6-lobed; lobes rounded to ovate or lanceolate; stamens free, anthers dehiscing longitudinally; rudimentary ovary absent. Female flowers: perianth 3–5-lobed; staminodes absent; ovary of 1 carpel; ovules solitary, pendulous, anatropous. Fruit fleshy, ovoid or ellipsoid, crowned by the persistent stigma. Seeds compressed-ellipsoidal.

A monotypic genus occurring in tropical and South Africa.

Xymalos monospora (Harv.) Warb. in Engl. & Prantl, Pflanzenfam. III, **6a**: 53, fig. 21/A, B (1893). —Oliver in Hooker's Icon. Pl. 25, t. 2444 (1896). —Perkins & Gilg in Engler, Pflanzenr. [IV, fam. 101]: 23, fig. 4/G–L (1901). —Sim, For. & For. Fl. Col. Cape Good Hope: 288, t. 121 (1907). —Baker & C.H. Wright in F.T.A. **6**, 1: 169 (1909). —Perkins in Engler, Pflanzenr. [IV, fam. 101] Nachtr: 10, fig 5 (1911). —C.H. Wright in F. C. **5**, 1: 493 (1912). —Eyles in Trans. Roy. Soc. South Africa **5**: 354 (1916). —Brenan, Check-list For. Trees Shrubs Tang. Terr.: 349 (1949). —J. Léonard in F.C.B. **2**: 402, t. 39 (1951). —Eggeling in F.W.T.A. ed. 2, **1**: 55 (1954). —F. White, F.F.N.R.: 56, fig. 12 (1962). —Verdcourt, F.T.E.A., Monimiaceae: 1 (1968). —Drummond in Kirkia **10**: 237 (1975). —K. Coates Palgrave, Trees Southern Africa: 174–175, figs. page 175 (1977). —Beentje, Kenya Trees Shrubs Lianas: 55 (1994). TAB. **13**. Syntypes from South Africa.

Xylosma? monospora Harv., Thes. Cap. **2**: 52, t. 181 (1863).
Toxicodendron acutifolium Benth. in J. Linn. Soc., Bot. **17**: 214 (1878). Type from South Africa.
Paxiodendron usambarense Engl., Pflanzenw. Ost-Afrikas C: 182 (1895). Type from Tanzania.
Paxiodendron usambarense var. *serratifolia* Engl., Pflanzenw. Ost-Afrikas C: 182 (1895). Syntypes from Tanzania.
Paxiodendron ulugurense Engl., Bot. Jahrb. Syst. **28**: 389 (1900). Type from Tanzania.
Xymalos usambarensis (Engl.) Engl., Bot. Jahrb. Syst. **30**: 310 (1901). Type as for *P. usambarense*.
Xymalos ulugurensis (Engl.) Engl., Bot. Jahrb. Syst. **30**: 310 (1901). —Baker & C.H. Wright in F.T.A. **6**, 1: 170 (1909). —Perkins in Pflanzenr. [IV, fam. 101] Nachtr: 11 (1911). Type as for *P. ulugurense*.

Evergreen shrubs or trees 3–20 m tall; branches glabrous; bark light to dark grey or yellowish-brown, flaking in large scales to leave conspicuous concentric ridged markings. Leaves 4–20 × 1.5–10 cm, elliptic to obovate, more rarely rotund, rounded to acute or shortly acuminate at the apex, cuneate at the base, entire to irregularly and coarsely glandular serrate on the margins, minutely pellucid-punctate, glabrous, slightly aromatic when crushed; lateral veins clearly looping well within the leaf margin, ± sunken above and prominent below giving the leaves a quilted appearance at least when fresh; petioles 0.3–3 cm long, glabrous. Inflorescences 1–5(7) cm long, racemose or paniculate, solitary or paired in the leaf axils, yellowish- or whitish-tomentose; bracts 1–2.5 mm long, triangular, white- to yellowish-tomentose. Flower unisexual, greenish, perianth 1–2 mm long; plants dioecious. Male flowers: perianth 4–6-lobed, lobes rounded to ovate; stamens 6–15(or more); anthers subsessile. Female flowers: perianth 3–5-lobed, lobes ovate, rounded or triangular; ovary obovoid, cylindrical or turbinate; stigma sessile, discoid, thick. Fruit 0.5–1.5(2.5) × 0.3–1.0(1.5) cm, ellipsoid, with persistent stigma at the apex, slightly asymmetric, fleshy, glabrous, orange or red when ripe. Seed solitary, white.

Zimbabwe. E: Chipinge Distr., Chirinda Forest, c. 1067 m, fl. viii.1966, *Goldsmith* 61/66 (K; LISC; PRE; SRGH). **Malawi**. N: Chitipa Distr., Misuku Hills, Mughesse rainforest, c. 1830 m, fl. 17.ix.1975, *Pawek* 10128 (K; MO; PRE; SRGH). S: Zomba Plateau, below summit on western edge, c. 1900 m, fl. 21.viii.1972, *Brummitt* 12962 (K; SRGH). **Mozambique**. N: Ribáuè, Serra de Mepáluè, c. 1600 m, young fr. 9.xii.1967, *Torre & Correia* 16406 (LISC). Z: Gurué, summit of Serra do Gurué, west of the Picos Namuli, near source of the Malema R., c. 1700 m, young fr. 4.i.1968, *Torre & Correia* 16921 (LISC). MS: Tsetserra, SE slopes, beneath villa of Carvalho, 1850 m, fl. 7.vi.1971, *Müller & Gordon* 1827 (LISC; SRGH).
   Also in the Cameroons and Bioko, and the Sudan southwards through Uganda, Kenya, Tanzania and eastern Zaire to South Africa. A scattered to locally frequent subcanopy tree of submontane evergreen rain forests, also in low altitude evergreen rain forest; 550–3000 m.
   The "quilted" leaves and the conspicuous rings rectangles and whorls, exposed when patches of bark flake off, make this species easily recognisable in the field.

Tab. 13. XYMALOS MONOSPORA. 1, flowering twig (×$\frac{2}{3}$), from *Holst* 4249; 2, leaf with coarsely serrate margins (×$\frac{2}{3}$), from *Chapman* 1643; 3, male inflorescence (×4); 4, male flower in section (×6), 3 & 4 from *Goldsmith* 61/66; 5, female flower (×6), from *Mavi* 436; 6, infructescence (×$\frac{2}{3}$), from *Pawek* 13025. Drawn by Eleanor Catherine.

145. LAURACEAE

By M.A. Diniz

Evergreen monoecious, or dioecious polygamous trees or shrubs, rarely leafless twining parasitic herbs with haustoria (*Cassytha*), all parts usually with aromatic oil glands. Leaves alternate, rarely opposite or subopposite, entire, coriaceous, rarely membranaceous, pinnately nerved or 3–5-nerved from near the base, usually net-veined, leaves reduced to small scales in *Cassytha*; stipules lacking. Inflorescences of cymes or paniculately arranged cymes, racemes or heads; flowers rarely solitary, axillary, usually subterminal; bracts simple, involucrate, deciduous; bracteoles present or lacking. Flowers hermaphrodite, unisexual by abortion, greenish, yellowish or reddish, actinomorphic. Receptacle campanulate, cup-shaped or urceolate, rarely oblong, sometimes accrescent after flowering. Perianth (2)3-merous in 2 whorls, not differentiated into calyx and corolla; tepals (4)6(8) free or ± connate, all similar or the outer ones smaller than the inner, imbricate or valvate in bud, deciduous or persistent. Stamens usually in 4 whorls inserted in the throat of the receptacle, or at the base of and opposite to the tepals, usually the fourth whorl, and sometimes the third whorl staminodal; filaments usually present and free, ± flattened, varying from very short to much longer than the anthers, those of the inner whorls mostly with a pair of globose glands at the sides or at the base (less often the glands attached to the receptacle, or sometimes the outer stamens with glands at the base); anthers basifixed, 2–4-celled, rarely 1-celled by confluence, cells ± collateral, or superposed (in 4-celled anthers), dehiscing from the base upwards; staminodes tepaloid, sagittate or ligulate. Ovary superior, rarely inferior (in Africa, *Hypodaphnis*), 1-locular; ovule solitary, anatropous from near the apex; style terminal; stigma small, discoid, pointed or unilaterally widened, rarely 2–3-lobed. Fruit baccate or drupaceous, indehiscent, free or ± surrounded by the accrescent receptacle, or completely enclosed within it, often borne on a thickened fruiting pedicel. Seed 1; testa membranaceous to coriaceous, sometimes adnate to the pericarp and indistinct; endosperm lacking; embryo straight; radicle superior, cotyledons fleshy, sometimes connate.

A large family with c. 50 genera and about 1000 species distributed mostly throughout the tropical and subtropical regions. Only 5 genera are native in Africa. From this predominantly woody family a few species are widely cultivated either as ornamentals, fruit trees or as a source of cinnamon and food flavouring. In the Flora Zambesiaca area species of *Cinnamomum*, *Laurus* and *Persea* are grown in gardens, or as street trees. R. Fouillay in Fl. Cameroun **18**: 78 (1974) and Verdcourt in F.T.E.A., Lauraceae: 2 (1996) provide keys to these genera.

Persea americana Mill. (avocado pear), native of tropical America, of which many varieties are grown in the Flora Zambesiaca area for its fruit. It is an evergreen tree 3–20 m high; leaves up to 20 × 12 cm, lanceolate, elliptic or obovate, obtuse to acute or acuminate at the apex, cuneate to rounded at the base; petiole up to 3.2 cm long. Flowers green in raceme-like inflorescences toward the end of the branches. Fruit large and pear-shaped, green to purplish when ripe, smooth-skinned to very rugose; seed ± globose, 4–6 cm in diameter.

Laurus nobilis L. (laurel or bay tree), from the Mediterranean, is grown as an ornamental tree but mainly for its leaves which are used for seasoning food. It is an evergreen up to c. 18 m high; leaves 4–10 × 1.5–6 cm, elliptic to oblong-lanceolate, obtuse or acute at the apex, cuneate or rounded at the base, pinnately veined, glabrous, entire, very aromatic. Flowers unisexual, small, yellowish-green, in axillary umbel-like inflorescences subtended by small involucres. Fruits black, 1–1.8 × 0.8–1 cm, ovoid-ellipsoid.

Cinnamomum camphora (L.) J. Presl (the camphor-tree), native of eastern Asia, is usually grown as an ornamental tree in the Flora Zambesiaca area. Natural camphor is prepared from the bark. It is a much-branched tree 6–12 m high; leaves mostly prominently 3-nerved from the base, 5–11 × 2–6 cm, ovate to ovate-elliptic, acuminate at the apex, cuneate at the base with glands in the axils of the primary nerves. Flowers small, yellow or white, arranged in panicles. Fruit small, black, fleshy.

Cinnamomum verum J. Presl (*C. zeylanicum* Blume), (cinnamon), native of eastern Asia, is widely cultivated in the tropics for its bark from which the true cinnamon of commerce is derived.

1. Twining parasitic herbs; leaves reduced to minute scales · · · · · · · · · · · · · · 1. **Cassytha**
 – Shrubs or trees; leaves well developed · 2
2. Anthers 4-celled · 4. **Ocotea**
 – Anthers 2-celled · 3

3. Pedicels with 3 bracteoles; mature ovary completely enclosed within the accrescent
   receptacle · 2. **Cryptocarya**
– Pedicels without bracteoles; mature ovary not enclosed within the receptacle · · · · · · · ·
   · **3. Beilschmiedia**

1. CASSYTHA L.

Cassytha L., Sp. Pl. 1: 35 (1753); Gen. Pl., ed. 5: 22 (1754). —Weber in J. Adelaide Bot. Gard.
3: 187–262 (1981).

Twining, parasitic herbs; stems filiform with small uniseriate haustoria. Leaves
reduced to minute ovate or lanceolate spirally arranged scales. Inflorescences
racemose, spicate or capitate, few-flowered. Flowers hermaphrodite, sessile or
subsessile in axils of small bracts and subtended by 2 bracteoles. Receptacle cup-
shaped, shallow, accrescent, completely enclosing the fruit. Tepals 6, in 2 whorls,
persistent, unequal, the outer 3 very small and bract-like, the inner 3 larger, valvate.
Stamens 9 in 3 whorls; the first and second whorls (outer whorls) with anthers
dehiscing introrsely, the second whorl occasionally staminodal, the third whorl with
anthers dehiscing extrorsely and with 2 glands attached to either side of the filament
base, the fourth (innermost) whorl staminodal; all the anthers 2-celled, splitting
longitudinally into 2 valves. Ovary ± sunken within the receptacle. Fruit drupaceous,
completely enclosed within the accrescent fleshy receptacle, and usually crowned by
the persistent tepals. Seed 1, with a coriaceous testa; cotyledons fleshy, distinct when
young, later confluent; embryo with a vertical axis.

A genus of about 17 species, mostly confined to Australia. One species is widely distributed
throughout tropical and subtropical regions of the world. Only 2 species are known in the
Flora Zambesiaca area.

Inflorescence a lax spike of up to 11 flowers, glabrous or puberulous; inner tepals glabrous
   inside; receptacle glabrous · 1. *filiformis*
Inflorescence subcapitate or very shortly spicate, (1)5-flowered, rusty- or fulvous-tomentose;
   inner tepals pubescent inside; receptacle rusty- or fulvous-pubescent   · · · · · 2. *pondoensis*

1. **Cassytha filiformis** L., Sp. Pl. 1: 35 (1753). —Meisner in De Candolle, Prodr. **15**, 1: 255
   (1864). —Baker, Fl. Mauritius: 292 (1877). —Engler, Pflanzenw. Ost-Afrikas C: 182 (1895).
   —Stapf in F.T.A. **6**, 1: 188 (1909); in F.C. **5**, 1: 500 (1912). —Engler, Pflanzenw. Afrikas
   (Veg. Erde 9) **3**, 1: 223, fig. 143 (1915). —Eyles in Trans. Roy. Soc. South Africa **5**: 354
   (1916). —Robyns & R. Wilczek in F.C.B. **2**: 443, t. 43 (1951). —Wild in Clark, Victoria Falls
   Handb.: 143 (1952). —Brenan in Mem. New York Bot. Gard. **9**: 60 (1954). —Friedrich-
   Holzhammer in Merxmüller, Prodr. Fl. SW. Africa, fam. 35 (1968). —Compton, Fl.
   Swaziland: 210 (1976). —Verdcourt in F.T.E.A., Lauraceae: 15 (1996). Type from India.
      Cassytha guineensis Thonn. ex Schumach., Beskr. Guin. Pl.: 199 (1827), as "*Cassyta*". Type
   from W Africa.
      Cassytha guineensis var. *livingstonii* Meisn. in De Candolle, Prodr. **15**, 1: 256 (1864).
   Syntypes: Mozambique, banks of Luabo R., *Kirk* 24 & 51 (K).

Dextrorsely twining herbs forming masses of brownish or greenish-yellow threads
over low vegetation; stems up to 2 m long, filiform, striate, branched, glabrous to
tomentose, green, bright brown or bright yellow. Leaves scale-like, 1.5–2 mm long,
ovate to lanceolate, apex acute. Inflorescence an axillary spike 1–6 cm long, usually
solitary, 3–10-flowered; peduncle 1–3 cm long; bracts and bracteoles c. 0.6 mm long,
ovate, ciliate. Receptacle glabrous. Flowers greenish-white, sessile, 1.5–2 mm long,
glabrous; the 3 outer tepals 0.6–1 mm wide, broadly ovate-orbicular and ciliate, the
3 inner tepals 1.8–2.4 × 1.5–2 mm, ovate-triangular, obtuse at the apex, glabrous
inside and outside, ± fleshy. Stamens 9, arranged in 3 whorls with an innermost or
fourth whorl made up of staminodes; filaments of the first and second whorls c. 0.3
mm long, anthers c. 1 × 0.6 mm, dehiscing introrsely; filaments of the third whorl c.
0.2 mm long with 2 globose glands at the base, anthers c. 0.9 × 0.3 mm, dehiscing
extrorsely; staminodes c. 0.4 mm long, triangular, glabrous, fleshy. Ovary 0.3 mm
long, ovoid; style 0.3 mm long. Fruit drupaceous, c. 6 × 5 mm, globose, surrounded
by the glabrous accrescent receptacle, crowned with a persistent perianth.

Botswana. N: Okavango R., N of Sepupa (Sepopa), fl. 4.v.1975, *Biegel, Müller & Gibbs Russell* 5091 (K; SRGH). **Zambia**. N: Mporokoso Distr., near Kalungwishi R., 1050 m, fl. & fr. 26.iv.1957, *Richards* 9436 (K). W: Ndola, fl. & fr. 14.vii.1932, *Young* 76 (BM; COI). C: 16 km N of Kabwe (Broken Hill), near Mulungushi R., fl. & fr. 11.v.1957, *Lister* 99 (SRGH). E: Katete, St. Francis' Hospital, c. 1000 m, fr. 25.v.1955, *Wright* 1 (K). **Zimbabwe**. N: Gokwe Distr., c. 22.5 km W of Gokwe on the Charama Road, fl. 26.iii.1964, *Bingham* 1235 (SRGH). W: Victoria Falls, 900 m, fl. 9.ii.1912, *Rogers* 5596 (K; SRGH). C: Ruwa Distr., Tanglewood Farm, 1350 m, v.1958, *Miller* 5305 (SRGH). E: Mutare Distr., Nyamakari R., Burma Valley, 750 m, fl. 22.ii.1962, *Chase* 7634 (BM; K). S: Save–Runde (Sabi–Lundi) junction, Chitsa's Kraal, 250 m, fl. 4.vi.1950, *Wild* 3361 (K; LISC; SRGH). **Malawi**. N: Karonga Distr., Vinthukutu Forest, c. 3 km N of Chilumba, 550 m, fl. & fr. 13.iv.1976, *Pawek* 10976 (K; MAL; MO). C: Dowa Distr., Lake Nyasa Hotel, 450 m, fl. & fr. 26.vii.1951, *Chase* 3893 (BM; COI; SRGH). S: Mulanje Distr., Mchese Mt., near Phalombe (Palombe), 750 m, fr. 18.vi.1958, *J.D. Chapman* H/666 (K; SRGH). **Mozambique**. N: Nampula Prov., Meconta, 16 km from Corrane to Liupo, c. 150 m, fl. & fr. 28.iii.1964, *Torre & Paiva* 11415 (J; LISC; LMA; MO; SRGH). Z: Pebane, 1 km on road from Mualama to Naburi, c. 100 m, fl. 15.i.1968, *Torre & Correia* 17158 (LISC; LMU; SRGH). T: Angónia Distr., from Ulónguè to Tete, between M'Salaázi R. and Namapanga R., fl. 14.ii.1980, *Macuácua, Stefanesco & Mateus* 1004 (LMA). MS: Manica Prov., S Chimanimani Mts., on the edge of the Haroni/Makurupini Forest, N bank of Makurupini R., fl. 11.vi.1971, *Pope* 479 (K; LISC). GI: Gaza Prov., between Manjacaze and Chongoéne, Missão de S. Benedito dos Muchopes, Mangunze, fl. & fr. 2.iv.1959, *Barbosa & Lemos* 8470 (K; LISC; LMA; SRGH). M: Maputo Prov., Manhiça, Macandzene, fl. & fr. 15.viii.1980, *Nuvunga & Mafumo* 271 (BM; SRGH).

Pantropical. Coastal vegetation usually on sandy dunes, margins of evergreen, gully and riverine forest, deciduous plateau and coastal woodlands, montane grasslands and dambo margins, and in vegetation of sandy beaches of lakes. On small bushes, woody shrubs and low trees, and on grasses and reeds. Recorded host plants in the Flora Zambesiaca area include species of *Baphia*, *Bauhinia*, *Crotalaria*, *Dalbergia*, *Indigofera*, *Kotschya*, *Tephrosia*, *Triumfetta*, *Ozoroa*, *Philippia*, *Xylopia*, *Euphorbia milii*, *Justicia*, *Phragmites*, *Miscanthidium* and *Chloris*; 0–1400 m.

2. **Cassytha pondoensis** Engl., Bot. Jahrb. Syst. **26**: 392 (1899). —Stapf in F.C. **5**, 1: 501 (1912). —Verdcourt in F.T.E.A., Lauraceae: 17 (1996). Syntypes from South Africa (Cape Province).

    Cassytha rubiginosa E. Mey. in Drège, Zwei Pflanzengeogr. Dokum.: 154, 171 (1843) nom. nud.

    Cassytha pubescens E. Mey. in Drège, Zwei Pflanzengeogr. Dokum.: 154 (1843) nom. nud. non R. Br.

Dextrorsely twining, parasitic herbs; stems very long, filiform, branched, pubescent to hairy, ± glabrescent; young stems rusty- or fulvous-tomentose. Leaves scale-like, 2–3 mm long, ovate, subobtuse at the apex, ± hairy. Inflorescence axillary, shortly pedunculate, subcapitate, or in short, 1–5-flowered spikes 0.5–2.5 cm long, rusty- or fulvous-tomentose; peduncles solitary or 2–4 together, up to 1 mm long; bracts and bracteoles 1–2 mm long, broadly ovate to suborbicular, subhyaline, rusty-pubescent and ciliate. Receptacle glabrous inside. Flowers yellowish or greenish, sessile, 2.5–3.5 mm long, 1–2 mm in diameter, rusty- to tawny-pubescent; the 3 outer tepals 1–1.5 mm long, suborbicular or ovate-lanceolate, obtuse or acute at the apex, ciliate, glabrous inside; the 3 inner tepals 1.5–2 mm long, lanceolate-triangular, subacute at the apex, pubescent inside. Stamens 9, arranged in 3 whorls with an innermost whorl of staminodes; filaments of the first whorl c. 1 mm long and broadly linear, ciliate, anthers c. 1 × 0.7 mm, dehiscing introrsely; stamens of the second whorl like the first, but smaller; filaments of the third whorl c. 0.9 mm long with 2 globose glands at the base, the anthers c. 0.6 × 0.4 mm dehiscing extrorsely; staminodes c. 0.3 mm long, ovate-triangular, sessile, glabrous, fleshy. Ovary c. 1 mm long, obovoid; style c. 1 mm long, filiform. Fruit drupaceous, 5–7 × 4–6 mm, globose or ellipsoidal, surrounded by the enlarged glabrous or ± puberulous receptacle, crowned at the apex by the persistent perianth.

Outer tepals suborbicular to broadly ovate, with apex obtuse; fruiting receptacle glabrous when mature · var. *pondoensis*
Outer tepals ovate-lanceolate, with apex acute; fruiting receptacle puberulous to pubescent when mature · var. *schliebenii*

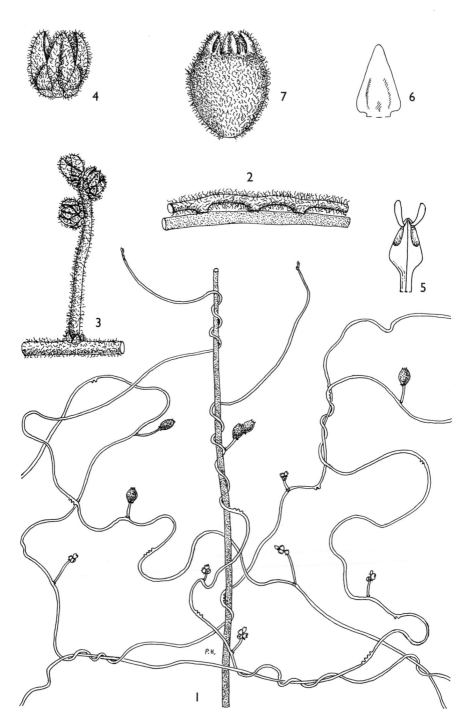

Tab. 14. CASSYTHA PONDOENSIS var. SCHLIEBENII. 1, habit (×²/₃); 2, detail of plant attached to host by haustoria; 3, inflorescence (×3); 4, flower (×8); 5, stamen (×12); 6, staminode (×12); 7, fruit (×4); 1–7 from *Macêdo* 3101. Drawn by Pat Halliday.

Var. **pondoensis**

Zambia. B: near Senanga, 1050 m, fl. & fr. 30.vii.1952, *Codd* 7232 (BM; K; SRGH). N: Mansa Distr., Samfya, fl. & fr. 30.i.1959, *Watmough* 204 (SRGH). W: Ndola, fl. & fr. 14.v.1954, *Fanshawe* 1197 (K; SRGH). S: Namwala, fl. 17.i.1961, *Mitchell* 4/89 (SRGH). **Zimbabwe**. E: Chimanimani Distr., Umvumvumvu R., 1350 m, fl. & fr. 7.vii.1955, *Chase* 5641 (BM; LMU; SRGH). **Malawi**. N: S of Nkhata Bay Village, near beach, fl. 6.ii.1977, *Grosvenor & Renz* 1055 (K; SRGH). S: Namasi, fl. & fr. 1897, *Cameron* s.n. (K). **Mozambique**. N: Lichinga (Vila Cabral), fl. 16.v.1934, *Torre* 100 (COI; LISC). Z: Gurué Mts., Lioma, fl. & fr. 1.vii.1943, *Torre* 5630 (BR; PRE; LMU; SRGH). GI: Gaza Prov., between Magul and Macia, fl. & fr. 1.vi.1959, *Barbosa & Lemos* 8553 (COI; K; LISC; LMU; SRGH). M: Matutuíne Distr., Salamanga, fl. & fr. 3.vii.1948, *Mendonça* 4511 (EA; LISC; LMU).

Also in Angola and South Africa. In open woodland, wooded grassland, riverine forest and coastal thicket; 0–1450 m.

Var. **schliebenii** (Robyns & R. Wilczek) M.A. Diniz, comb. et stat. nov. TAB. **14**. Type from Tanzania.

    Cassytha schliebenii Robyns & R. Wilczek in Bull. Jard. Bot. État **19**: 506 (1949). Type as above.

Differs from var. *pondoensis* in having fulvo-pubescent stems, the spikes shorter and few-flowered, the flowers fulvo-pubescent, the outer tepals ovate-lanceolate, the ovary ovoid, and the fruiting receptacle puberulous to pubescent when ripe.

Zambia. N: Mbala Distr., c. 11 km S of Mbala (Abercorn), fl. & fr. 21.vii.1930, *Hutchinson & Gillett* 4029 (BM; K; LISC; SRGH). **Malawi**. N: Mzimba Distr., Marymount, Mzuzu, 1350 m, fl. 23.v.1972, *Pawek* 5382 (SRGH). C: Ntchisi Mt., 1400 m, fl. & fr. 25.vii.1946, *Brass* 16945 (BM; K; SRGH). **Mozambique**. N: Mecubúri, Serra Chinga, between Chinga 1 and Chinga 2, 790 m, fl. & fr. 27.iv.1968, *Macêdo* 3101 (LMA).

Also in Tanzania. In dense forest and *Brachystegia* woodland; 780–1500 m.

Loveridge 1616 (LISC; SRGH) collected in Melsetter, and *Wild* 2879 (LISC; SRGH) collected in Chimanimani Mts. are intermediate between *C. filiformis* and *C. pondoensis*. The outer tepals of these specimens are glabrous outside and therefore approach *C. filiformis*, their inner tepals on the other hand have a few hairs near the base, and the filaments of the first whorl are ciliate at the base and so approach *C. pondoensis*.

2. CRYPTOCARYA R. Br.

Cryptocarya R. Br., Prodr. Fl. Nov. Holl. **1**: 402 (1810). —Bentham & Hooker, Gen. Pl. **3**: 150, excl. syn. (1880), nom. conserv.

Evergreen trees or shrubs. Leaves alternate, sometimes sub-opposite, pinnately-nerved or 3-nerved from near or at the base. Cymes paniculate, few- to many-flowered, axillary and subterminal. Flowers hermaphrodite; pedicels 3-bracteolate (in African species). Receptacle ± cylindrical, constricted in upper half, persistent. Tepals 6, subequal, erect or ± patent, caducous. Fertile stamens 9 in 3 whorls, with staminodes in the fourth (innermost) whorl; filaments very short (in African species); anthers 2-celled, those of the 2 outer whorls opening introrsely (in African species), the third (inner) whorl opening extrorsely (in African species); filaments of inner whorls with a pair of sessile or stalked glands at the base. Staminodes shortly stipitate, ovate, acute. Ovary sessile, ovoid or ellipsoid, slender, glabrous, enclosed in the receptacle; style cylindric, slender; stigma slightly expanded. Fruit drupaceous, globose or ellipsoid, 1-seeded, completely enclosed within the hard or fleshy accrescent receptacle with only a minute orifice at the apex; pericarp membranaceous or crustaceous, and can usually be separated from the receptacle. Seed globose; testa membranaceous.

A genus of over 200 species in the tropics and subtropics, mostly from South America, the Indo-Malayan region and Australia. Six species occur in South Africa with 3–4 extending into East Africa. In Madagascar there are c. 8 endemic species.

1. Leaves ovate, conspicuously 3–nerved from above the base with 3–4 lateral veins arising from the midrib, rounded or obtuse at the apex · 1. *liebertiana*
– Leaves elliptic, all nerves pinnate (or lower pair making a more acute angle with midrib and running ± parallel to leaf margin from or just above the leaf base), acute to obtuse and apiculate or cuspidate-acuminate at the apex · 2

2. Leaves broadly elliptic to suborbicular, obtusely cuspidate-acuminate and apiculate at the apex · 2. *woodii*
- Leaves mostly narrowly elliptic, sometimes more or less oblong, acute to acuminate at the apex · 3. *transvaalensis*

1. **Cryptocarya liebertiana** Engl., Bot. Jahrb. Syst. **26**: 390, t. 10 fig. B (1899). —Burtt Davy in Bull. Misc. Inform., Kew **1908**: 156 (1908), as "*Siebertiana*". —Stapf in F.T.A. **6**, 1: 172 (1909). —Engler, Pflanzenw. Afrikas (Veg. Erde 9) **3**, 1: 223 (1915). —Kostermans in Bull. Jard. Bot. État **15**: 102 (1938) pro parte exclud. syn. *Cryptocarya transvaalensis.* —Brenan, Checklist For. Trees Shrubs Tang. Terr.: 262 (1949). —Palmer & Pitman, Trees of Southern Africa **1**: 601, cum 3 fig. (p. 599) & 1 photogr. (p. 600) (1972) pro parte exclud. syn. *C. transvaalensis.* —Dowsett-Lemaire in Bull. Jard. Bot. Nat. Belg. **55**: 383 (1985). —Beentje, Kenya Trees Shrubs Lianas: 56 (1994). —Verdcourt in F.T.E.A., Lauraceae: 7, t. 2 (1996). Syntypes from Tanzania.

Medium to large evergreen tree (8)10–35(50) m high, sometimes flowering as a shrub c. 5 m high; stem buttressed; crown ± rounded; bark brown to grey-brown, smooth with shallow longitudinal fissures and flaking in small patches; slash brown, fine-grained. Young branches and apical buds densely rusty-tomentellous, soon glabrescent, older twigs with smooth or longitudinally lenticellate grey bark. Leaves leathery; petiole 5–12 mm long, slightly channelled above, glabrous; lamina (3)4–10(11) × (1.7)2–4.5(6.5) cm, broadly ovate or ovate-elliptic, obtuse to rounded and ± emarginate at the apex, rarely obtusely and shortly acuminate, broadly cuneate at the base, ± undulate on the margin; at first rusty-tomentellous, particularly on the nerves, soon glabrescent on both surfaces, except sometimes along the midrib and the nerves beneath; lateral nerves in 4–5 pairs, the lowest nerves opposite or sub-opposite arising above the leaf base, obliquely to the midrib and remote from the next nerves, reticulation obscure; upper surface glossy dark green, with midrib slightly impressed; lower surface greenish or cinnamon-brown, becoming bluish-green with prominent midrib and nerves, reticulation very dense, obscure. Cymes paniculately arranged, few-flowered, axillary; rhachis 1.5–5 cm long, usually fulvo-tomentellous; peduncle 0.5–2.5 cm long; bracts lanceolate, minute, caducous. Flowers white to cream-coloured, small, cinnamon-tomentellous; pedicels up to 1 mm long; receptacle 1–1.5 mm long, cylindric-urceolate, constricted near the apex, tomentellous outside and inside; tepals erect-patent, 2–2.5 mm long, elliptic, acute and incurved at apex, tomentellous outside and inside. Fertile stamens 9 in 3 whorls, with a fourth (innermost) whorl of staminodes; filaments of outer whorl tomentellous, shorter than the tepals, anthers c. 1 mm long, acute at the apex; filaments of inner whorls similar but shorter, glands c. 0.5 mm in diameter, globose with a pilose stalk; staminodes c. 1 mm long with short pilose filaments and ovate-lanceolate acute glabrous antherodes. Ovary up to 1 mm long, ovoid, glabrous; style short, stigma capitate. Fruit 1.3–1.8 cm in diameter, globose or nearly so, black and somewhat fleshy outside; mature ovary enveloped by the accrescent receptacle; fruiting pedicel thick, 2–4 mm long.

Zimbabwe. E: Nyanga Distr., between Circular Drive and Pungwe Gorge, 1750 m, st. 7.xii.1966, *Müller* 554 (K). **Malawi. N:** Nkhata Bay Distr., 8 km east of Mzuzu at Roseveare's Cottage, 1200 m, fr. 21.iv.1973, *Pawek* 6565 (K). **S:** Zomba Plateau, Mlunguzi Stream near Mandala Falls, st. *J.D. Chapman* 1443 (SRGH). **Mozambique. N:** Ribáuè, Serra Mepáluè, 1300 m, fl. 5.xii.1967, *Torre & Correia* 16367 (COI; FHO; LISC; LMA; PRE). **Z:** Serra Gurué, 3 km from Licungo Falls, 1200 m, fr. immat. 24.ii.1966, *Torre & Correia* 14852 (BR; J; LISC; LMU; SRGH). **MS:** Báruè, Serra de Choa, 12 km from Catandica (Vila Gouveia) to Posto Administrativo de Choa, 1300 m, fr. 26.v.1971, *Torre & Correia* 18670 (COI; EA; LISC; LMA; LMU; SRGH).

From Tanzania to South Africa (KwaZulu-Natal). In mixed evergreen forest, evergreen rainforest and submontane gully forest; 700–2000 m.

In its typical form this species has ovate leaves strongly 3-nerved from above the base. However, leaf shape and nervation is somewhat variable and *C. liebertiana*, as here circumscribed, exhibits forms which approach *C. latifolia* and *C. transvaalensis* in leaf characters. Further work may reveal that *C. liebertiana* should be recognized as subspecies of *C. latifolia.*

Cryptocarya latifolia Sond. is a species from coastal Eastern Cape Province and KwaZulu-Natal in South Africa, and is distinguished from *C. liebertiana* by its leaves being more rounded at the apex and broadly elliptic or obovate, with margins more revolute and with the lower pair of lateral nerves arising closer to the leaf base and running half way or more to the leaf apex. The inflorescence of *C. latifolia* is usually more densely ferruginous velvety than in *C. liebertiana.* Specimens *Torre & Correia* 16367 (LISC) and 18670 (LISC) from Ribáuè and Báruè respectively, are forms of *C. liebertiana* with broadly elliptic leaves rounded at the apices.

The ground bark mixed with crocodile fat is used by the Zulus to treat chest troubles. The root is used as a purgative in Tanzania. The wood is white and soft.

2. **Cryptocarya woodii** Engl., Bot. Jahrb. Syst. **26**: 391 (1899). —Stapf in F.C. **5**, 1: 496 (1912). —Bews, Fl. Natal Zululand: 94 (1921). —Kostermans in Bull. Jard. Bot. État **15**: 99, pl. 8 (1938). —Palmer & Pitman, Trees of Southern Africa **1**: 602 cum 1 fig. (p. 602) and 2 photogr. (p. 600) (1972). —K. Coates Palgrave, Trees Southern Africa, ed. 3 revised: 180 (1990). Type from South Africa (KwaZulu-Natal).

Cryptocarya acuminata Sim, For. Fl. Col. Cape Good Hope: 289 (err. t. CXXIII), t. 158 (1907); For. Fl. Port. E. Africa: 95 (1909). Type from South Africa.

An evergreen shrub, or small to medium sized tree 4–10 m high, occasionally up to 20 m. Bark smooth, grey or greyish-brown; branches terete, glabrous with ± prominent lenticels, becoming verruculose; young branchlets subangular and sparsely appressed pubescent particularly towards the apex; apical buds sericeous-tomentellous. Leaves chartaceous to subcoriaceous, glabrous; petiole 3–5 mm long; lamina 1.5–9 × 1.5–5 cm, ± orbicular to elliptic, cuspidate-acuminate or abruptly acuminate with a blunt tip, rounded to ± cuneate at the base, slightly revolute and undulate on the margin; upper surface drying greyish-brown with an impressed midrib and 3–4(7) pairs of slightly raised lateral nerves; lower surface with midrib and lateral nerves prominent; tertiary veins finely and inconspicuously reticulate. Cymes paniculately arranged, few-flowered, axillary; rhachis 1–3.5 cm long, sparsely and minutely appressed pubescent; peduncle 0.2–1 cm long, slender; bracts ovate-lanceolate, minute, soon caducous. Flowers c. 3 mm long, greenish-white to yellowish, pedicels 1–5 mm long, slender; receptacle 1–1.5 mm long, subcylindric, enlarged near the apex, sparsely appressed tomentellous outside, glabrous inside; tepals erect-patent, 1–1.5 mm long, ovate-elliptic, acute and incurved at the apex. Fertile stamens in 3 whorls, with an innermost whorl of staminodes; filaments very short, with the inner slightly shorter than the outer, slender, pilose; anthers up to 0.8 mm long, ovate, apiculate, glabrous; glands c. 0.4 mm in diameter, subglobose with stalks c. 0.2 mm long, pilose; staminodes c. 1 mm long with pilose filaments and ovate, acute to apiculate, glabrous antherodes. Ovary 0.5–0.8 mm long, ovoid to ellipsoid, glabrous; style 1.5–2 mm long, stigma capitate. Fruit 1.5–2 cm in diameter when ripe, subglobose, mature ovary enveloped by the dark purplish-black receptacle; fruiting pedicel slender.

Mozambique. M: Matutuíne, fl. & fr. 23.ii.1947, *R.M. Hornby* 2544 (K; LMA; SRGH).
Also in Swaziland and South Africa (from eastern Transvaal and southwards to eastern Cape Province). In coastal forests in the Flora Zambesiaca area, in riverine forest, woodland and in evergreen forest margins in South Africa; up to 1500 m in South Africa.
The wood is brown, close grained and hard.

3. **Cryptocarya transvaalensis** Burtt Davy in Bull. Misc. Inform., Kew **1922**: 334 (1922). TAB. **15**. Type from South Africa (Transvaal).
Cryptocarya libertiana sensu auct. incl. Kostermans in Bull. Jard. Bot. État **15**: 102 (1938) pro parte quoad syn. *C. transvaalensis*. —Palmer & Pitman, Trees of Southern Africa **1**: 601 (1972) pro parte quoad syn. *C. transvaalensis*. —Drummond in Kirkia **10**: 237 (1975).

An evergreen tree 6–15(25) m high; trunk up to 25 cm in diameter; crown rounded; bark thin, ± smooth, dull dark brown or grey; slash brown. Apical buds ferruginous-tomentose, soon glabrescent; branchlets ± angular, puberulous; older twigs terete, glabrous. Leaves coriaceous, green above, glaucescent beneath; petiole 4–10 mm long, channelled above, puberulous, soon glabrous; lamina (3)6–10(12) × (1.5)3–4.5 cm, narrowly elliptic, rarely oblong, acute-acuminate or shortly cuspidate-acuminate with an obtuse tip, ± broadly cuneate or rounded at the base, undulate on the margins; lateral nerves 4–7, ± alternate on each side of the midrib, the lowermost sometimes oblique, opposite or subopposite, ascending and arching near the margin; upper surface glossy, dark green, glabrous, with midrib and lateral nerves impressed; lower surface dull bluish-green, glabrous except on the midrib when young, midrib prominent and lateral nerves slightly raised; tertiary venation hardly visible, reticulate, slightly raised. Cymes paniculately arranged, densely many-flowered, axillary, borne mostly towards the ends of the branches; rhachis 1–7.5 cm long, sparsely and minutely appressed-pubescent; peduncle 1–2 cm long. Flowers

Tab. 15. CRYPTOCARYA TRANSVAALENSIS. 1, flowering twig (×²/₃), from *Ball* 3 and *Chase* 515; 2, flower (×8); 3, longitudinal section through flower (×6), 2 & 3 from *Chase* 515; 4, stamens of inner and outer whorls (×8); 5, stamens plus stalked glands arising from base of inner whorl stamen (×8); 6, staminode, innermost whorl (×8); 7, fruit (×1). Drawn by Pat Halliday.

greenish-yellow, 2.5–3.5 mm long; pedicels up to 2 mm long; receptacle 1.0–1.5 mm long, urceolate-cylindric, sparsely and minutely appressed-pubescent outside, densely pubescent inside; tepals erect, 1–2 mm long, elliptic or ovate-elliptic, obtuse-acute at the apex, ± appressed-pubescent outside and inside. Fertile stamens included; filaments short, the inner ones slightly shorter than the others, all pilose; anthers c. 0.6 mm long, ovate, connective conspicuously pointed, glabrous; glands c. 0.5 mm in diameter, subglobose; staminodes c. 1 mm long with short pilose filaments and ovate, acute antherodes. Ovary c. 1 mm long, ellipsoid, glabrous; style 1.0–1.5 mm long; stigma minutely capitate. Fruit 1.3–1.5 cm in diameter when ripe, subglobose, mature ovary enveloped by the fleshy, plum-red to purple accrescent receptacle, glabrous; pericarp thin.

Zimbabwe. E: Chipinge Distr., Chirinda Forest, c. 1100 m, fl. & fr. v.1962, *Goldsmith* 132/62 (COI; K; LISC; LMA; SRGH). **Mozambique**. MS: Chimoio Distr., Serra Garuso, fr. 5.iv.1948, *Andrada* 1109 (COI; EA; LISC; LMU; MO; PRE; SRGH).
Extending southwards from eastern Zimbabwe to the Transvaal in South Africa. Canopy tree of montane and submontane evergreen forest; 950–1200 m.
The fruits are eaten by birds and monkeys.

3. BEILSCHMIEDIA Nees

Beilschmiedia Nees in Wallich, Pl. Asiat. Rar. **2**: 69 (1831). —Robyns & R. Wilczek in Bull. Jard. Bot. État **19**: 459–506 (1949); **20**: 197–226 (1950). —Rohwer in Kubitzki et al., Fam. Gen. Vasc. Pl. **2**: 385 (1993).
Tylostemon Engl., Bot. Jahrb. Syst. **26**: 389 (1899).
Afrodaphne Stapf in J. Linn. Soc., Bot. **37**: 110 (1905).

Trees or shrubs. Leaves alternate, occasionally subopposite towards ends of branches, subsessile or petiolate, papyraceous to coriaceous, pinnately nerved. Inflorescences terminal and axillary, panicles or racemes or paniculate cymes, few- to many-flowered; bracts ± pubescent, caducous. Flowers hermaphrodite; receptacle shortly tubular or cup-shaped. Tepals 6(8) in 2 whorls on rim of receptacle, subequal, deciduous. Stamens in 4 whorls, the 2 outer whorls each of 3(4) fertile stamens with anthers dehiscing introrsely; the inner (third) whorl with anthers dehiscing extrorsely or laterally and with a gland on either side at the base of the filaments, or inner (third) whorl anthers sometimes sterile; the innermost (fourth) whorl reduced to 3 staminodes. Ovary subglobose to narrowly ellipsoid, sessile, ± immersed in the receptacle; narrowed above into a slender style with an obscure stigma. Fruit a 1-seeded drupe, subglobose to ovoid or ellipsoid, naked.

A pantropical genus of about 200 species with c. 80 species occurring in Africa and Madagascar. Robyns & R. Wilczek (1950) recognize 77 species in tropical continental Africa. However, Verdcourt in F.T.E.A. (1966) suggests that there are probably fewer than this, pointing out that only 2(3) species are recorded from East Africa. Two species are represented in the Flora Zambesiaca area.

Fruit c. 3 × 1.5 cm; leaves usually ± rounded-obtuse at the apex, sometimes with an obscurely acuminate tip, lateral nerves in 8–14 pairs; outer stamens ± sessile; stamens of third whorl with short filaments and with 2 reniform glands attached to the receptacle near the base of the filaments; staminodes 3, comprising the innermost whorl · 1. *ugandensis* var. *katangensis*
Fruit 3.5–6 × 2–2.5 cm; leaves often more tapered, the apex usually acute to somewhat acuminate, lateral nerves in 6–10 pairs; outer stamens with distinct filaments (0.6 mm long); stamens (staminodes) of the third whorl with 2 globose glands attached to the filaments; staminodes 3–6, arranged in the inner 2 whorls · · · · · · · · 2. *gilbertii* var. *glabra*

1. **Beilschmiedia ugandensis** Rendle in J. Linn. Soc., Bot. **37**: 203 (1905). —Robyns & R. Wilczek in F.C.B. **2**: 421 (1951). —F. White, F.F.N.R.: 432 (1962). —Verdcourt in F.T.E.A., Lauraceae: 4 (1996). Type from Uganda.
Tylostemon ugandensis (Rendle) Stapf in F.T.A. **6**, 1: 181 (1909). Type as above.

A shrub or small tree up to 12(15) m high; bole 20–30 cm in diameter; bark greyish, rough; young branches brownish, drying blackish, angular, smooth, puberulous to glabrous; buds fulvous appressed-pubescent. Leaves subcoriaceous or coriaceous, glabrous, drying dull brown above, paler beneath; petiole, 0.6–1.2 cm long; lamina 6.0–21.5 × 2.5–8.0 cm, oblong-elliptic to elliptic or obovate, obtuse or shortly acuminate at the apex, cuneate to rounded at the base; midrib flat above, strongly prominent below, lateral nerves in 8–14 pairs, tertiary nerves forming a fine reticulum visible on both surfaces but prominent below. Inflorescence a terminal or axillary panicle of cymes 3–9 cm long, many-flowered, laxly much-branched, rufous-pubescent or greyish-puberulous all over; peduncle 0.5–2.0 cm long. Flowers reddish or greenish-brown, 1.5–2.5 mm long and 3–4 mm in diameter at anthesis; pedicels up to 1.5 mm long; receptacle ± two-thirds the length of the flower, cup-shaped, puberulous within; tepals ± one-third the length of the flower, rounded-ovate, obtuse at the incurved apex, puberulous inside and outside. Fertile stamens 9, glabrous, arranged in 3 whorls; stamens of the first and the second (outer) whorls 0.5–0.8 × 0.5–1.0 mm, sessile or subsessile, hairy at the base with anthers dehiscing introrsely; stamens of the third whorl with short hairy filaments, anthers dehiscing laterally to extrorsely with 2 reniform glandular appendages attached to the receptacle near the base of the filaments; staminodes 3, in a fourth whorl, c. 0.5 mm long, lanceolate-triangular, subsessile, pubescent. Ovary c. 0.7 mm long, ovoid, glabrous, immersed in the receptacle; style c. 0.7 mm long with a pointed stigma. Drupe pinkish or purplish, ellipsoid or ovoid, mature ovary not enveloped by the receptacle. Embryo axis vertical.

Var. **katangensis** Robyns & R. Wilczek in Bull. Jard. Bot. État **19**: 473 (1949); in F.C.B. **2**: 422 (1951). Type from Zaire.

Small tree. Leaves coriaceous. Inflorescence 3–9 cm long, minutely appressed greyish puberulous. Flowers 2–2.5 mm long, 3–4 mm in diameter. Stamens of the first and second whorls c. 0.8 × 1.0 mm, glabrescent. Drupe c. 3.0 × 1.5 cm, ellipsoid, dark purplish.

Zambia. N: Kawambwa, fl. 15.xi.1957, *Fanshawe* 4036 (FHO; K). W: Mwinilunga, fr. 10.ix.1955, *Holmes* 1193 (K).
Also in Zaire and Central African Republic. In mushitu swamp and riverine forest; c. 1200 m.
This variety differs from the var. *ugandensis* by its longer, puberulous panicles and its longer flowers. Var. *ugandensis* occurs in Central Africa and Uganda, in riverine forest.

2. **Beilschmiedia gilbertii** Robyns & R. Wilczek in Bull. Jard. Bot. État **19**: 504 (1949); in F.C.B. **2**: 440 (1951). Type from Zaire.
    Beilschmiedia sp. 1 (*White* 3331, FHO; K) F. White, F.F.N.R.: 58 (1962).

A shrub or small tree up to 6 m high; bark green to greyish, smooth; young branches greyish or dark reddish, striate, glabrous. Leaves papyraceous to thinly coriaceous, glabrous, drying greenish above, pale brown beneath; petiole 0.6–1.3 cm long, canaliculate above, rough; lamina (6)13–24 × (2.3)4–8 cm, oblong-lanceolate to obovate-oblong, acute to shortly acuminate at the apex, cuneate at the base; slightly revolute and undulate on the margins; midrib and nerves impressed above, prominent below, venation finely reticulate; lateral nerves in 6–9 pairs. Inflorescence a panicle of cymes in the axils of subterminal leaves; panicles 4–15 cm long, lax, many-flowered, whitish puberulous; peduncles 0.5–2 cm long. Flowers pinkish to reddish, 2–3 mm long and 2–3 mm in diameter at anthesis, appressed-puberulous; pedicels 1–4 mm long; receptacle c. 1 mm long, cylindrical to ± campanulate, puberulous outside, glabrous inside; tepals c. 1 mm long, broadly ovate, obtuse or rounded at the apex, puberulous, strongly ciliate. Stamens 6(8), pubescent, arranged in 2 whorls; filaments c. 0.6 × 0.6 mm, pubescent; anthers 0.2 mm long dehiscing extrorsely. Staminodes 6 in 2 whorls, (whorls 3 and 4); those in the third whorl c. 1 mm long, cylindrical, sometimes 1 or 2 are fertile with anthers dehiscing laterally and with 2 globose glands at the base; those in the fourth whorl 0.5–0.8 mm long and 0.3–0.6 mm wide at the base, ± triangular to cordiform, puberulous outside, glabrescent inside. Ovary 0.8–1.0 mm long, globose, puberulous towards the apex, style 0.5–1.0 mm long. Drupe brownish-purple, 3.5–6.0 × 2.0–2.5 cm, ellipsoid or fusiform and curved, mucronate at the apex. Mature ovary not enveloped by receptacle. Seed ± cylindrical.

Tab. 16. BEILSCHMIEDIA GILBERTII var. GLABRA. 1, flowering twig (×²/₃), from *Milne-Redhead* 2950A; 2, longitudinal section through flower, diagrammatic; 3, fertile stamen of outer whorl (×18); 4, sterile stamen of third whorl with globose glands at base of filament (×18); 5, staminode, inner whorl (×18), 3–5 from *Holmes* 1149; 6, fruit (×²/₃); 7, seeds (×²/₃), 6 & 7 from *Milne-Redhead* 2950. Drawn by Pat Halliday.

Var. **glabra** Robyns & R. Wilczek in Bull. Jard. Bot. État **19**: 506 (1949); in F.C.B. **2**: 442 (1951). TAB. **16**. Type from Zaire.

Shrub or small tree 4–6 m high. Inflorescences 4–9 cm long. Flowers 2–3 mm long; stamens of the 2 outer whorls c. 1 mm long, 0.7–0.9 mm wide; staminodes of the fourth whorl c. 0.8 mm long, 0.6 mm wide near the base. Ovary c. 1 mm long, ovoid, glabrous; style up to 1 mm long. Drupe brownish-purple, 3.5–6.0 × 2.0–2.5 cm, ellipsoid or fusiform and curved, mucronate at the apex. Seed ± cylindrical. Embryo axis transverse.

Zambia. W: Mwinilunga Distr., Matonchi R., fl. 18.ii.1938, *Milne-Redhead* 2950A (BM; K; LISC). Also in Zaire. In riverine or dry evergreen forest; c. 1000 m.
This variety is distinguished by its glabrous ovary. Var. *gilbertii* is rare, being restricted to Forestier Central in Zaire where it occurs in evergreen forest.

4. OCOTEA Aubl.

Ocotea Aubl., Hist. Pl. Guiane Fr. **2**: 780 (1775). —Bentham & Hooker f., Gen. Pl. **3**: 157 (1880). —Rohwer in Mitt. Inst. Allg. Bot. Hamburg **20**: 1–278 (1986).

Evergreen trees or shrubs. Leaves alternate, rarely sub-opposite, simple, pinnately-nerved, sometimes also 3-nerved near the base, membranous to coriaceous, glabrous or pubescent. Inflorescence of cymules arranged in axillary or subterminal panicles. Flowers hermaphrodite or unisexual; receptacle broadly obconic, enlarging in fruit to form a truncate or 6-lobed cupule. Tepals 6, subequal, usually spreading. Hermaphrodite flowers: fertile stamens 9 in 3 whorls adnate to the tepals, those of the third whorl with lateral glandular appendages at the base, filaments short; staminodes 3 in a fourth (innermost) whorl situated on the receptacle; ovary ovoid, usually glabrous. Male flowers: similar but ovary sterile, stalk-like or lacking. Female flowers: as in hermaphrodite flowers but stamens rudimentary. Fruit drupaceous, partially enveloped by the accrescent receptacle and acorn-like in appearance in some species.

A large genus of about 200 species, mostly in tropical and subtropical America, with c. 50 species in tropical and subtropical Africa, Madagascar, the Comoro Islands, the Mascarenes, Madeira and Canary Islands.
The timber of *Ocotea* spp. is used in construction of buildings, *O. bullata* (Burch.) Baill., the Cape Stinkwood, being much prized for furniture making. This species and *O. kenyensis* are declared protected plants in South Africa. *O. bullata* may be distinguished by its bullate leaves (blisters in the axils of the lower nerves, corresponding with hair-lined pit-like domatia on the leaf lower surface), and by its fruit c. 2 × 0.85 cm in size.

Mature leaves discolorous; lower surface silvery-white-pruinose, usually ± sparsely puberulous at least on the main veins; lateral nerves in 6–8(9) pairs; fruit c. 8–11 × 6–6.5 mm, borne in a cup 4–6 mm wide and 2–3 mm long · 1. *usambarensis*
Mature leaves concolorous, greenish and glabrous or subglabrous on both surfaces, not pruinose on lower surface; lateral nerves in 8–10 pairs; fruit c. 9–27 × 7–14 mm, borne in a cup 10–12 mm wide and 7–10 mm long · 2. *kenyensis*

1. **Ocotea usambarensis** Engl. in Abh. Königl. Akad. Wiss. Berlin **1894**, 1: 51, 54 (1894); Pflanzenw. Ost-Afrikas **C**: 182 (1895). —Stapf in F.T.A. **6**, 1: 187 (1909). —Engler, Pflanzenw. Afrikas (Veg. Erde 9) **1**, 1: 302, fig. 268 (1910). —Mildbraed, Wiss. Ergeben. Deutsch. Zentr.-Afr.-Exped., Bot., part 3: 215 (1911). —Kostermans in Bull. Jard. Bot. État **15**: 85 (1938). —Brenan, Check-list For. Trees Shrubs Tang. Terr.: 262 (1949). —Robyns & R. Wilczek in F.C.B. **2**: 405 (1951). —Topham, Check List For. Trees Shrubs Nyasaland Prot.: 58 (1958). —Dale & Greenway, Kenya Trees & Shrubs: 243 (1961). —F. White, F.F.N.R.: 58 (1962). —Dowsett-Lemaire in Bull. Jard. Bot. Belg. **55**: 383 (1985). —Beentje, Kenya Trees Shrubs Lianas: 56 (1994). —Verdcourt in F.T.E.A., Lauraceae: 10 (1996). TAB. **17**. Type from Tanzania.

A medium to large evergreen tree 15–30(40) m high, with a dense crown and a stout straight bole up to 10 m high, buttressed at the base. Bark purplish-grey, rough when old and flaking in small rounded or rectangular scales; wood

Tab. 17. OCOTEA USAMBARENSIS. 1, flowering twig (×²/₃); 2, flower, with reference to
Hooker's Icones Pl. **30**: t. 2934 (1911) (×8); 3, stamen of outer staminal whorl (×8); 4,
stamen of third whorl, external view, with 2 glands at base of filament (×8), 1–4 from
Greenway 5521; 5, fruits (×²/₃), from *Angus* 874. Drawn by Pat Halliday.

aromatic with a camphor-like scent when freshly cut, slash a light reddish-brown, becoming darker with exposure. Young branches and twigs slender, angular and finely longitudinally striate towards the ends, ± densely fulvous-pubescent or tomentose, sometimes glabrescent. Leaves aromatic with a camphor-like scent, membranous to coriaceous, discolorous drying brownish above and silvery-white or greyish on the lower surface; petiole 0.5–2.0 cm long, tomentose to glabrescent; lamina (2.5)4.5–14(17) × (2)2.5–6(7.5) cm, ovate to elliptic, acuminate or sometimes acute or obtuse at the apex, rounded to truncate rarely subcordate or cuneate at the base, entire; upper surface glossy dark green and ± puberulous especially on the midrib and nerves, or glabrous; lower surface silvery-white pruinose and ± puberulous, at least on the midrib and nerves, rarely glabrescent; lateral nerves in 6–8(9) pairs, slightly raised above, prominent beneath; tertiary nerves net-veined, ± inconspicuous above. Inflorescence a panicle of cymes in the axils of subterminal leaves, panicles 3–13 cm long, ± lax, few to many-flowered, densely yellowish-brown pubescent; peduncle 2–8 cm long; bracts minute, ovate to ovate-lanceolate, soon falling. Flowers small, yellowish, densely pubescent; pedicels 1–4 mm long; receptacle 1 mm long, broadly obconical, glabrous within; tepals 2–3 mm long, elliptic to ovate-elliptic, obtuse to rounded at the apex, pubescent within. Stamens 9 in 3 whorls, with staminodes forming a fourth (innermost) whorl; filaments of the first and second (the outer) whorls short and broad, flattened, their anthers 1.0–1.5 mm long, elliptic-rectangular, obtuse at the apex and dehiscing introrsely; filaments of the third whorl ± square, pilose, with 2 shortly-stalked globose to ovate glands, the anthers oblong, a little longer than the others, truncate or emarginate, dehiscing extrorsely; staminodes (in the innermost whorl) c. 1 mm long, ovate-triangular, sessile or subsessile. Ovary 1.0–1.5 mm long, ovoid, glabrous; style short thick glabrous; stigma subpeltate. Fruit orange-brownish, up to 10 × 6 mm, ellipsoid, the lower one third enclosed in the accrescent ± fleshy receptacle.

Zambia. N: Muchinga Escarpment, on road to Mpika, c. 48 km S of Ishiba Ngandu (Shiwa Ngandu), fl. & fr. immat. 29.xi.1952, *Angus* 874 (BM; COI; K). **Malawi**. N: Mwenembwe Forest, W side of upper Henga Valley, c. 32 km from Livingstonia, 1800 m, st. 2.xii.1952, *J.D. Chapman* 59 (BM; K).
   Also from Zaire, Uganda, Kenya and Tanzania. Canopy tree, often an emergent, in submontane and montane evergreen rainforests, and infrequent in evergreen swamp forest; up to 2350 m.
   The decoction of the decorticated bark is used in Tanzania for the relief of stomach ache.

2. **Ocotea kenyensis** (Chiov.) Robyns & R. Wilczek in F.C.B. **2**: 406, pl. 40 (1951). —Dale & Greenway, Kenya Trees & Shrubs: 242 (1961). —Palmer & Pitman, Trees of Southern Africa **1**: 595, cum photogr. & fig. (1972). —R. Wilczek & Troupin in Fl. Rwanda **1**: 265 (1978). —Dowsett-Lemaire in Bull. Jard. Bot. Belg. **55**: 383 (1985). —K. Coates Palgrave, Trees Southern Africa, ed. 3, revised: 176 (1990). —Beentje, Kenya Trees Shrubs Lianas: 56 (1994). —Verdcourt in F.T.E.A., Lauraceae: 11 (1996). Type from Kenya.
   Ocotea gardneri Hutch. & M.B. Moss in Bull. Misc. Inform., Kew **1930**: 70, fig. on page 69 (1930) non Mez (1889). Type from Kenya.
   Tylostemon kenyensis Chiov., Racc. Bot. Miss. Consol. Kenya: 107 (1935). Type as above.
   Ocotea gardneri var. *cuneata* Lebrun, Ess. For. Rég. Mont. Congo Orient.: 79 (1935). Type from Zaire.
   Ocotea viridis Kosterm. in Bull. Jard. Bot. État **15**: 83 (1938). Type as for *Ocotea gardneri*.

An evergreen tree 6–20(40) m high; bole straight, unbranched for c. 8 m, 1–1.5 m in diameter; bark brownish or reddish-grey, becoming rough and longitudinally scaling; underbark bright orange, slash cream quickly turning blackish-grey. Branches terete, rough with conspicuous leaf scars; young branches and twigs ± angular, densely whitish or greyish-puberulous, soon glabrescent. Leaves aromatic, coriaceous or subcoriaceous, drying blackish-brown or greyish-green, often somewhat paler beneath; petiole 1–1.8 cm long, thick, puberulous, soon glabrescent; lamina (5)8–20(22) × 3.5–9(10) cm, elliptic to broadly elliptic, ovate-elliptic or ovate, obtuse to rounded or acute at the apex, broadly cuneate or sometimes rounded at the base, glabrous except for a few minute appressed whitish hairs towards the base of the midrib beneath; midrib and nerves slightly prominent beneath, lateral nerves in 8–10 pairs, tertiary venation visible on both surfaces but more conspicuous beneath. Inflorescences solitary or clustered in leaf axils at the ends of the branches, with flowers arranged in cymes or grouped in terminal panicles of cymes up to 7 cm long, ± lax, few-flowered, whitish puberulous or glabrescent; peduncle up to 2.5 cm long; bracts c. 1 mm long, somewhat obtusely triangular, soon falling. Flowers cream-coloured, fragrant, ± appressed-puberulous; pedicels 1–4 mm long, thickening in fruit, angular; receptacle 1–1.5 mm long, obconic, glabrous within; tepals up to 3.5 mm long, obtusely ovate, densely pubescent within, glandular-pubescent at the margins. Fertile stamens 9 in 3 whorls, with staminodes forming a fourth (innermost) whorl; stamens c. 2 mm long, glabrous or with some hairs at the filament base; filaments of the first and second whorls usually shorter than anthers, flat; anthers broadly ovate or ± angular, obtuse, truncate or somewhat emarginate at the apex, dehiscing introrsely; filaments of the third whorl c. 1 mm long with 2 sessile oblong glands at the base, each c. 1 mm long, anthers ± ellipsoidal and rounded at both ends, dehiscing extrorsely; staminodes 3, c. 1.7 mm long, lanceolate, dilated at the base, acute at the apex, or staminodes lacking. Ovary 1.0–1.5 mm long, ovoid, glabrous; style very short, glabrous; stigma large, densely covered in long papillae. Fruit up to 2.2 × 1.2 cm, ovoid-ellipsoid, with the lower one-third to one-half enclosed in the accrescent fleshy receptacle.

Mozambique. N: Nampula Prov., Serra Chinga 2, 1100 m, fr. immat. 29.v.1968, *Macêdo & Macuácua* 3300 (LMA). Z: Serra Gurué, near source of Malema R., 1700 m, fl. & fr. 5.i.1968, *Torre & Correia* 16955 (COI; EA; FHO; LISC; LMU; PRE; SRGH). MS: Sofala Prov., Serra de Gorongosa, SE slopes, c. 1100 m, st. 25.vii.1970, *Müller & Gordon* 1452 (K; LISC; LMA; SRGH).
   Also in Kenya, Tanzania, Zaire and eastern South Africa (Mpumalanga and KwaZulu-Natal). Canopy trees in medium to high altitude evergreen rain forest; 1100–2600 m.
   Müller 3737 from Zimbabwe is cultivated.
   Verdcourt, in F.T.E.A., Lauraceae: 13 (1996), notes that the taxonomy of *Ocotea kenyensis* is difficult and treats it as one variable species extending from Ethiopia to South Africa. He considers, however, that some East African populations may merit subspecific rank on the basis of their narrower leaves and fruit.

146. HERNANDIACEAE

By B.L. Stannard

   Trees, shrubs or lianes, with aromatic oils in stems and leaves. Leaves alternate, exstipulate, simple, digitately compound or 3–5-lobed. Inflorescences of axillary, rarely terminal, corymbose or paniculate cymes, pedunculate, bracteate or ebracteate. Flowers regular, bisexual, or unisexual by abortion and the plants monoecious or polygamous, rarely dioecious. Perianth segments subequal, free or shortly united below, in two 3–5-merous, valvate whorls or in one 4–8-merous, imbricate whorl. Stamens 3–5 in one whorl and opposite the outer segments when perianth is double, or (2)4(7) when perianth is single; anthers 2-thecous, dehiscing introrsely or laterally by 2 flap-like valves. Staminodes absent or ± gland-like, 1–2-whorled, usually outside the stamens; outer whorl paired at the base of the stamens, inner whorl alternating with the stamens. Ovary inferior, 1-locular; ovule solitary, pendulous, anatropous. Fruit dry indehiscent, often winged, longitudinally ribbed or with lateral wings (*Illigera*), or with 2 apical wings formed by accrescent perianth segments (*Gyrocarpus*), or wingless and enclosed in an inflated capsule (*Hernandia*). Seed solitary, without endosperm; embryo straight; cotyledons plano-convex, ± lobed, or flattened and spirally twisted.

   A family of 5 genera and some 60 species in both the Old and New World tropics. Only one genus, *Gyrocarpus*, is found in the Flora Zambesiaca area. This and *Sparattanthelium* have been treated by some authors as a separate family but this account follows Kubitzki's monograph (Bot. Jahrb. Syst. **89**: 78–209, figs. 1–51 (1969) Kubitzki et al., Fam. Gen. Vasc. Pl. **2**: 334 (1993)) in which these two genera are treated as a subfamily *Gyrocarpoideae*.

GYROCARPUS Jacq.

Gyrocarpus Jacq., Select. Stirp. Amer. Hist.: 282, t. 178, f. 80 (1763). —Bentham & Hooker f., Gen. Pl. **1**: 689 (1865). —Kubitzki et al., Fam. Gen. Vasc. Pl. **2**: 334 (1993).

Deciduous trees, rarely shrubs. Leaves simple, undivided or 3–5-lobed, long-petioled. Inflorescences of corymbose, ebracteate cymes, mainly in upper leaf axils. Flowers small, numerous, hermaphrodite and unisexual on the same individual, predominantly male. Perianth segments in one 4–8-merous imbricate whorl, equal or unequal with 2 segments larger and accrescent in hermaphrodite flowers. Stamens (2)4(7), exserted, inserted at base of perianth; anthers dehiscing laterally by upward-opening valves. Staminodes as many as and alternating with the stamens, or more numerous. Ovary and style reduced or absent in male flowers. Fruit a globose ovoid or ellipsoid drupe with 2 long apical wings formed by accrescent perianth segments. Cotyledons spirally twisted.

A genus of 3 species from the tropics of both the Old and New World. Only one species occurs in the Flora Zambesiaca area.

Gyrocarpus americanus Jacq., Select. Stirp. Amer. Hist.: 282, t. 178, f. 80 (1763). —Meisner in De Candolle, Prodr. **15**, 1: 247 (1864). —Engler, Pflanzenw. Ost-Afrikas **C**: 182 (1895). —Verdcourt in Kew Bull. **21**: 254 (1967). —Kubitzki in Bot. Jahrb. Syst. **89**: 182 (1969). —Verdcourt in F.T.E.A., Hernandiaceae: 7 (1985). —Beentje, Kenya Trees Shrubs Lianas: 66 (1994). Type from Colombia.

Subsp. **africanus** Kubitzki in Bot. Jahrb. Syst. **89**: 186 (1969). —Verdcourt in F.T.E.A., Hernandiaceae: 8 (1985). TAB. **18**. Type from Eritrea.
 Gyrocarpus americanus sensu auct. incl. —Engl., Pflanzenw. Ost-Afrikas **C**: 182 (1895) pro parte. —F. White, F.F.N.R.: 58 (1962). —Schreiber in Prodr. Fl. SW. Afrika, fam. 36: 1 (1968). —Palmer & Pitman, Trees of Southern Africa **1**: 605, figs. pages 117, 604, 606 (1972). —Jacobsen in Kirkia **9**: 157 (1973). —Drummond in Kirkia **10**: 237 (1975). —K. Coates Palgrave, Trees Southern Africa: 181, fig. 41 (1977), non Jacq. sensu stricto.
 Gyrocarpus asiaticus sensu Sprague in F.T.A. **6**, 1: 190 (1909), pro parte. —Hutchinson, Botanist Southern Africa: 308, fig. (1946), non Willd. sensu stricto.
 Gyrocarpus sp. (Allen 401), Eyles in Trans. Roy. Soc. South Africa **5**: 355 (1916).
 Gyrocarpus jacquinii sensu Brenan, Check-list For. Trees Shrubs Tang. Terr.: 245 (1949) pro parte non Gaertn. sensu stricto.

Deciduous trees 3–20 m tall, branching well above the ground and with an open rounded crown, usually flowering and fruiting when leafless. Stems smooth, pale grey or yellowish-brown to light silvery-grey, outer bark papery and easily rubbed off to show green underneath. Leaves clustered at the branch tips; blades soft textured, 4–15 × 3.5–14 cm, ovate, adult leaves ± deeply 3(5)-lobed, sometimes unlobed, acuminate at the apex, rounded and shortly tapering into the petiole, or broadly cuneate at the base, strongly 3(5)-veined from the base, pubescent to glabrescent and dark green on upper surface, greyish-green to whitish-grey tomentose on lower surface interspersed with longer hairs particularly along the veins; lobes up to half the blade in length, acute to acuminate and with rounded sinuses, median lobe broadly triangular or ± constricted towards the base; petioles up to 10 cm long, sulcate (in dried state at least), puberulous. Flowers small, in many dense clusters at the ends of the inflorescence branches, hermaphrodite and unisexual. Perianth c. 1 mm long, densely pubescent. Stamens 4, up to c. 3 mm long, shorter in hermaphrodite flowers; filaments usually pilose, sometimes with dorsal glands; anthers 0.75–1 mm long, glabrous. Staminodes alternating with the stamens, 0.5–1 mm long, clavate, pubescent. Style and stigma pubescent. Fruit pendulous, seed-bearing part hard, indehiscent, 1.3–2.0 × 1.0–1.5 cm, globose to ellipsoid, rugose (in dried state at least), glabrous, light green to reddish-brown; wings 5.5–9.0 × 0.7–1.6 cm, one usually slightly longer than the other, narrowly oblanceolate-spathulate, united at the base, pubescent towards the base, yellowish-green to dark brown.

Zambia. C: South Luangwa National Park (Luangwa Valley Game Res. South), c. 6.4 km S of Katete/Luangwa R. confluence, c. 610 m, fl. & fr. 27.iv.1966, *Astle* 4849 (K; SRGH). S: Livingstone Distr., Victoria Falls, W side of Palm Grove, fr. 29.viii.1947, *Brenan & Greenway* in

Tab. 18. GYROCARPUS AMERICANUS subsp. AFRICANUS. 1, twig with inflorescences and leaves at the apex ($\times^2/_3$), from *Wild* 7693; 2, fruits ($\times^2/_3$), from *Pereira & Correia* 2786; 3, leaf shapes ($\times^1/_6$), from *Wild* 7693, *Mullin* 92/56, *Mullin* 92/56; 4, flower ($\times$12), from *Wild* 7693. Drawn by Eleanor Catherine.

Brenan 7775 (K). **Zimbabwe**. N: Hurungwe Distr., Nyanyanya R., fr. 27.xi.1956, *Mullin* 92/56 (PRE; SRGH). W: in gorge below Victoria Falls, fr. s.d., *Allen* 401 (K; SRGH). E: Chipinge Distr., W end of Mwangazi Gap, st. 29.i.1975, *Pope, Biegel & Russell* 1437 (K; MO; PRE; SRGH). S: Ndanga, Triangle Sugar Estate, fl. iii.1949, *Bates* in *GHS* 23089 (K; SRGH). **Malawi**. S: Mangochi Distr., Monkey Bay, island in bay, fl. 29.ii.1968, *Wild* 7693 (K; SRGH). **Mozambique**. T: between Songo and Cahora Bassa dam, below marker 400, fl. 25.iii.1972, *Macêdo* 5097 (K; LMU; PRE; SRGH). MS: Sofala Prov., Gorongosa Nat. Park, "road 5 area", fl. & fr. 20.iv.1973, *Tinley* 2775 (K; LISC; MO; PRE; SRGH). GI: on banks of Limpopo R., 80 km below Pafuri, fr. vii.1939, *Smuts* 2194 (K; PRE).

Also in Eritrea, Tanzania, Angola, Namibia and South Africa. Occasional, at low to medium altitudes, usually in hot dry areas, on rocky ridges or stony slopes of escarpments and river valleys, in riverine thickets and deciduous woodlands; 50–1175 m.

Kubitzki (loc. cit.) recognizes eight related subspecies of *G. americanus* of which only subsp. *africanus* occurs the Flora Zambesiaca area. Subsp. *americanus* is widespread in both the Old and New World tropics but in Africa is restricted to Kenya and Tanzania. It can be distinguished from subsp. *africanus* by its unlobed adult leaves with their glabrous to fuscous pubescent lower surfaces. Subsp. *pinnatilobus* from West Africa has leaves with divided lobes. The other subspecies all occur outside mainland Africa. Subsp. *capuronianus*, subsp. *glaber* and subsp. *tomentosus* are apparently endemic to Madagascar and as a group are distinguished by having stamens with glabrous, as opposed to pilose, filaments. Subsp. *pachyphyllus* is restricted to the Australian mainland and subsp. *sphenopterus* to N Australia and possibly the Philippines and Lesser Sunda Islands.

Kubitzki points out that subsp. *africanus* shows a distinct clinal variation, with fruit size, leaf lobing and density of indumentum gradually decreasing towards the southern and south-western part of its distribution. Specimens of subsp. *africanus* with unlobed leaves can be distinguished from the other subspecies with unlobed leaves by the greyish-green to whitish-grey tomentum on the lower leaf surface.

INDEX TO BOTANICAL NAMES